European Women in Mathematics

Proceedings of the 13th General Meeting

European Women in Mathematics

Proceedings of the 13th General Meeting

University of Cambridge, UK   3 – 6 September 2007

Catherine Hobbs
Oxford Brookes University, UK

Sylvie Paycha
Université Blaise Pascal, France

editors

 **World Scientific**

NEW JERSEY · LONDON · SINGAPORE · BEIJING · SHANGHAI · HONG KONG · TAIPEI · CHENNAI

Published by

World Scientific Publishing Co. Pte. Ltd.

5 Toh Tuck Link, Singapore 596224

USA office: 27 Warren Street, Suite 401-402, Hackensack, NJ 07601

UK office: 57 Shelton Street, Covent Garden, London WC2H 9HE

British Library Cataloguing-in-Publication Data
A catalogue record for this book is available from the British Library.

EUROPEAN WOMEN IN MATHEMATICS
Proceedings of the 13th General Meeting

ISBN-13 978-981-4277-67-9
ISBN-10 981-4277-67-3

Printed in Singapore.

PREFACE

The *Thirteenth General Meeting of European Women in Mathematics* (EWM07) was held at the University of Cambridge, UK, 3–6 September, 2007. The present volume contains the texts of most of the invited talks delivered at the conference and a selection of the contributed papers. All have been peer-reviewed. The volume also contains some articles on women in mathematics, such as historical accounts, which were presented during the meeting.

The EWM meetings have been taking place since 1989. They feature prominent women mathematicians as speakers, and generally have two or three main mathematical themes as well as sessions on issues affecting women in mathematics. Many have been amazed and encouraged by the experience of attending an EWM conference, never having previously been part of a group of over 100 women listening intently to a talk on state-of-the-art mathematical research, or had the opportunity to meet and talk to women mathematicians in a variety of fields. The conferences have sparked collaborations, follow-on meetings on related themes and, most importantly, have inspired many women from graduate students to professors as they develop their careers as working mathematicians.

The conference was supported by grants from the London Mathematical Society, EPSRC, the University of Cambridge and Schlumberger. Springer and Google also supported the conference with prizes for best contributed paper.

The Local Organizing and Scientific Committees are to be thanked for creating a well-run and productive meeting, with an exciting programme of talks and poster presentations. We especially thank Ms. Amanda Stagg for all her administrative support before and during the conference.

S. Paycha   Université Blaise Pascal, France
C. A. Hobbs   Oxford Brookes University, UK
(Editors)   30 June 2009

ORGANIZING COMMITTEE

Eva Bayer Fluckiger	– Ecole Polytechnique Fédéral de Lausanne, Switzerland
Anne Davis (Chair)	– University of Cambridge, UK
Catherine Hobbs	– Oxford Brookes University, UK
Marjo Lipponen	– University of Turku, Finland
Ursula Martin	– Queen Mary University of London, UK
Sylvie Paycha	– Université Blaise Pascal, Clermont-Ferrand, France
Tsou Sheung Tsun	– University of Oxford, UK
Caroline Series	– University of Warwick, UK

LOCAL ORGANIZING COMMITTEE

Anne Davis (Chair)	– University of Cambridge, UK
Rachel Camina	– University of Cambridge, UK
Vivien Easson	– University of Cambridge, UK
Catherine Hobbs	– Oxford Brookes University, UK
Susan Pitts	– University of Cambridge, UK
Caroline Series	– University of Warwick, UK
Ursula Martin	– Queen Mary University of London, UK

CONTENTS

PART A

Invited Talks

DEFORMATION QUANTISATION AND CONNECTIONS

S. GUTT

Université Libre de Bruxelles
Campus Plaine, CP 218
bd du Triomphe
1050 Brussels, Belgium
and
Université P. Verlaine, Metz
Ile du Saulcy
57045 Metz Cedex 01, France
sgutt@ulb.ac.be

After a brief introduction to the concept of formal Deformation Quantisation, we shall focus on general konwn constructions of star products, enhancing links with linear connections.

We first consider the symplectic context: we recall how any natural star product on a symplectic manifold determines a unique symplectic connection and we recall Fedosov's construction which yields a star product, given a symplectic connection.

In the more general context, we consider universal star products, which are defined by bidifferential operators expressed by universal formulas for any choice of a linear torsionfree connection and of a Poisson structure. We recall how formality implies the existence (and classification) of star products on a Poisson manifold. We present Kontsevich formality on $\mathbb{R}^d$ and we recall how Cattaneo-Felder-Tomassini globalisation of this result proves the existence of a universal star product.

Keywords: Deformation quantisation; star product; symplectic connection; universal star product.

1. Quantization

Quantisation of a classical system is a way to pass from classical to quantum results.

Classical mechanics is considered in its Hamiltonian formulation on the motion space, which is the quotient of the evolution space (usually the product of the phase space and the real line) by the trajectories. Thus the framework is a symplectic manifold (or, more generally, when one deals with constraints, a Poisson manifold).

In this setting a point represents a motion, so that an observable is represented by a family of smooth functions on that manifold M. The dynamics is defined in terms of a Hamiltonian $H \in C^\infty(M)$ and the time evolution of an observable $f_t \in C^\infty(M \times \mathbb{R})$ is governed by the equation:

$$\frac{d}{dt} f_t = -\{H, f_t\}.$$

Quantum mechanics is considered in its usual Heisenberg's formulation. The framework is a Hilbert space (states are rays in that space). An observables is described by a one parameter family of selfadjoint operators on that Hilbert space. The dynamics is defined in terms of a Hamiltonian H, which is a selfadjoint operator, and the time evolution of an observable A_t is governed by the equation:

$$\frac{dA_t}{dt} = \frac{i}{\hbar}[H, A_t].$$

A natural suggestion for quantisation is a correspondence $Q\colon f \mapsto Q(f)$ mapping a function f to a self adjoint operator $Q(f)$ on a Hilbert space $\mathcal{H}$ in such a way that $Q(1) = \mathrm{Id}$ and

$$[Q(f), Q(g)] = i\hbar Q(\{f, g\}).$$

There is no such correspondence defined on all smooth functions on M when one puts an irreducibility requirement which is necessary not to violate Heisenberg's principle.

To deal with this problem, various mathematical theories of quantization were proposed. Deformation Quantisation was introduced in the seventies by Flato, Lichnerowicz and Sternheimer in [10] and developed in [3]; they "suggest that quantisation be understood as a deformation of the structure of the algebra of classical observables rather than a radical change in the nature of the observables."

One stresses here the fundamental aspect of the space of observables rather than the set of states; observables behave indeed in a nicer way when one deels with composed systems: both in the classical and in the quantum picture, the space of observables for combined systems is the tensor product of the spaces of observables.

The algebraic structure of classical observables that one deforms is the algebraic structure of the space of smooth functions on a Poisson manifold: the associative structure given by the usual product of functions and the Lie structure given by the Poisson bracket. Formal deformation quantisation is defined in terms of a formal deformation of that structure called a star product.

2. Basic definitions

Definition 2.1. A **Poisson bracket** defined on the space of smooth functions on a manifold M, is a $\mathbb{R}$- bilinear skewsymmetric map:

$$C^\infty(M) \times C^\infty(M) \to C^\infty(M) \quad (u,v) \mapsto \{u,v\}$$

satisfying Jacobi's identity ($\{\{u,v\},w\} + \{\{v,w\},u\} + \{\{w,u\},v\} = 0$) and Leibniz rule ($\{u,vw\} = \{u,v\}w + \{u,w\}v \quad \forall u,v,w \in C^\infty(M)$).

Leibniz rule says that bracketing with a given function u is a derivation of the associative algebra of smooth functions on M, hence is given by a vector field X_u on M. By skewsymmetry, a Poisson bracket is thus given in terms of a contravariant skew symmetric 2-tensor P on M, called the **Poisson tensor**, by

$$\{u,v\} = P(du \wedge dv). \tag{1}$$

The Jacobi identity for the Poisson bracket Lie algebra is equivalent to the vanishing of the Schouten bracket:

$$[P,P] = 0.$$

(The Schouten bracket is the extension -as a graded derivation for the exterior product- of the bracket of vector fields to skewsymmetric contravariant tensor fields; it will be developed further in section 4.1.)

A **Poisson manifold**, denoted (M,P), is a manifold M with a Poisson bracket defined by the Poisson tensor P.

A particular class of Poisson manifolds, essential in classical mechanics, is the class of symplectic manifolds.

Definition 2.2. A **symplectic manifold**, denoted by (M,ω), is a manifold M endowed with is a closed nondegenerate 2-form ω. The corresponding Poisson bracket of two functions $u,v \in C^\infty(M)$ is defined by

$$\{u,v\} := X_u(v) = \omega(X_v, X_u), \tag{2}$$

where X_u denotes the **Hamiltonian vector field** corresponding to the function u, i.e. such that $i(X_u)\omega = du$. In coordinates, the components of the corresponding Poisson tensor P^{ij} form the inverse matrix of the components ω_{ij} of ω.

Examples of symplectic manifolds are given by **cotangent bundles**; if N is a manifold, its cotangent bundle $T^*N \xrightarrow{\pi} N$ is endowed with a canonical 1-form λ defined by

$$\lambda_\xi(Y) = \langle \xi, \pi_* Y \rangle \qquad \xi \in T^*N, \ Y \in T_\xi(T^*N),$$

where $\langle \cdot, \cdot \rangle$ denotes the pairing between the tangent space $T_x N$ at a point $x = \pi(\xi) \in N$ and its dual space, the cotangent space at x, $T_x^* N$. Then $(T^* N, d\lambda)$ is a symplectic manifold.

Duals of Lie algebras form the class of linear Poisson manifolds. If $\mathfrak{g}$ is a Lie algebra then its dual $\mathfrak{g}^*$ is endowed with the Poisson tensor P defined by

$$P_\xi(X, Y) := \xi([X, Y])$$

for $X, Y \in \mathfrak{g} = (\mathfrak{g}^*)^* \sim (T_\xi \mathfrak{g}^*)^*$.

Definition 2.3. [3] A **star product** on (M, P) is a bilinear map

$$C^\infty(M) \times C^\infty(M) \ \to C^\infty(M)[\![\nu]\!] \qquad (u, v) \mapsto u * v = \sum_{r \geq 0} \nu^r C_r(u, v)$$

such that

(1) extending the map $\mathbb{R}[\![\nu]\!]$-bilinearly to $C^\infty(M)[\![\nu]\!] \times C^\infty(M)[\![\nu]\!]$, it is formally associative:

$$(u * v) * w = u * (v * w);$$

(2) (a) $C_0(u, v) = uv$, (b) $C_1(u, v) - C_1(v, u) = \{u, v\}$;
(3) $1 * u = u * 1 = u$.

When the C_r's are bidifferential operators on M, one speaks of a **differential star product**. When, furthermore, each bidifferential operator C_r is of order maximum r in each argument, one speaks of a **natural star product**.

Given any star product $*$ on (M, P) and any series $T = \mathrm{Id} + \sum_{r=1}^{\infty} \nu^r T_r$ of linear operators on $C^\infty(M)$, one can build a new star product $*'$ defined by $f *' g = T(T^{-1} f * T^{-1} g)$. Remark that any such series T can be written in the form $T = \mathrm{Exp}\, E$ where $E = \sum_{r=1}^{\infty} \nu^r E_r$ with the E_r linear operators on $C^\infty(M)$ and where Exp denotes the exponential series. This motivates the following definition of equivalence:

Definition 2.4. Two star products $*$ and $*'$ on (M, P) are **equivalent** if and only if there is a series

$$E = \sum_{r=1}^{\infty} \nu^r E_r$$

where the E_r are linear operators on $C^\infty(M)$, such that

$$f *' g = \mathrm{Exp}\, E\, ((\mathrm{Exp} - E)\, f * (\mathrm{Exp} - E)\, g)). \tag{3}$$

Remark 2.1. Two differential star products $*$ and $*'$ on (M, P) are equivalent iff there is a series $E = \sum_{r=1}^{\infty} \nu^r E_r$ where the E_r are differential operators , giving the equivalence [i.e. such that $f *' g = \operatorname{Exp} E \left((\operatorname{Exp} - E) f * (\operatorname{Exp} - E) g \right)$].

Two natural star products $*$ and $*'$ on (M, P) are equivalent iff there is a series $E = \sum_{r=1}^{\infty} \nu^r E_r$ where the E_r are differential operators of order at most $r + 1$, giving the equivalence.

An example: Moyal star product on $\mathbb{R}^n$

Consider a vector space $V = \mathbb{R}^m$ with a Poisson structure P with constant coefficients:

$$P = \sum_{i,j} P^{ij} \partial_i \wedge \partial_j, \quad P^{ij} = -P^{ji} \in \mathbb{R}$$

where $\partial_i = \partial/\partial x^i$ is the partial derivative in the direction of the coordinate x^i, $i = 1, \ldots, n$. **Moyal star product** is defined by

$$(u *_M v)(z) = \exp \left(\frac{\nu}{2} \sum_{r,s} P^{rs} \partial_{x^r} \partial_{y^s} \right) (u(x)v(y)) \Bigg|_{x=y=z} . \qquad (4)$$

Associativity follows from the fact that

$$\partial_{t^k} (u *_M v)(t) = (\partial_{x^k} + \partial_{y^k}) \exp \left(\frac{\nu}{2} P^{rs} \partial_{x^r} \partial_{y^s} \right) (u(x)v(y)) \Bigg|_{x=y=t} ,$$

so that

$$((u *_M v) *_M w)(x') = \exp \left(\frac{\nu}{2} P^{rs} \partial_{t^r} \partial_{z^s} \right) ((u *_M v)(t)w(z)) \Bigg|_{t=z=x'}$$

$$= \exp \left(\frac{\nu}{2} P^{rs} (\partial_{x^r} + \partial_{y^r}) \partial_{z^s} \right) \exp \left(\frac{\nu}{2} P^{ab} \partial_{x^a} \partial_{y^b} \right) (u(x)v(y))w(z)) \Bigg|_{x=y=z=x'}$$

$$= \exp \left(\frac{\nu}{2} P^{rs} (\partial_{x^r} \partial_{z^s} + \partial_{y^r} \partial_{z^s} + \partial_{x^r} \partial_{y^s}) \right) (u(x)v(y))w(z)) \Bigg|_{x=y=z=x'}$$

$$= (u *_M (v *_M w)(x').$$

Remark that one can define by an analogous formula a Moyal star product on a Poisson manifold (M, P) as soon as the Poisson tensor writes

$$P = \sum_{i,j} P^{ij} X_i \wedge X_j, \quad P^{ij} = -P^{ji} \in \mathbb{R}$$

where the X_i's form a set of commuting vector fields on M.

Definition 2.5. When the constant Poisson structure P is non degenerate (which implies $V = \mathbb{R}^{2n}$), the space of polynomials in ν whose coefficients are polynomials on V with Moyal product is called **the Weyl algebra** $(S(V^*)[\nu], *_M)$.

Moyal star product on $(\mathbb{R}^{2n}, \omega = \sum_i dp_i \wedge dq_i)$ is related to the composition of operators via Weyl's quantisation. This associates to a polynomial f on $\mathbb{R}^{2n}$ a differential operator $Q_W(f)$ on $\mathbb{R}^n$ in the following way: to the classical observables q^i and p_i, one associates the quantum operators $Q^i = q^i\cdot$ and $P_i = -i\hbar\frac{\partial}{\partial q^i}$ acting on functions depending on q^j's. Since Q^i and P_j do no longer commute, one has to specify which operator is associated to a higher degree polynomial in q^i and p_j. The *Weyl ordering* associates the corresponding totally symmetrized polynomial in Q^i and P_j, e.g.

$$Q_W(q^1(p^1)^2) = \frac{1}{3}(Q^1(P^1)^2 + P^1Q^1P^1 + (P^1)^2Q^1).$$

Then

$$Q_W(f) \circ Q_W(g) = Q_W(f *_M g) \quad \text{(for } \nu = i\hbar\text{).}$$

Remark that another ordering, such as the standard ordering, which associates to a polynomial f in q^i and p_j the operator $Q_{std}(f)$ with all the Q^i's on the left and the P^j's on the right, gives another isomorphism between the space of differential operators on $\mathbb{R}^n$ and the space of polynomials on $\mathbb{R}^{2n}$. This yields another star product $*_{std}$ on $\mathbb{R}^{2n}$ so that

$$Q_{std}(f) \circ Q_{std}(g) = Q_{std}(f *_{std} g) \quad \text{(for } \nu = i\hbar\text{).}$$

One can show that $Q_W(f) = Q_{std}(\exp \frac{i\hbar}{2} Df)$ where $D = \frac{\partial^2}{\partial p \partial q}$ so that

$$\exp \nu D(f *_M g) = \exp \nu D(f) *_{std} \exp \nu D(g).$$

One can show more generally that any two differential star products on $\mathbb{R}^{2n}$ are equivalent.

3. Symplectic case: star products and symplectic connections

A linear connection on a manifold M is a way to differentiate a vector field along a vector field:

Definition 3.1. Let $\chi(M) = \Gamma(TM)$ be the space of smooth vector fields on M. A **linear connection** on M is a bilinear map

$$\nabla : \chi(M) \times \chi(M) \to \chi(M) \quad (X,Y) \mapsto \nabla_X Y$$

such that $\nabla_{fX}Y = f\nabla_X Y$ and $\nabla_X fY = X(f)Y + f\nabla_X Y, \forall X, Y \in \chi(M)$ and $\forall f \in C^\infty(M)$. Equivalently, it is a $C^\infty(M)$-linear map

$$\nabla : \Gamma(TM) \to \Gamma(T^*M \otimes TM) \quad Y \mapsto \nabla Y$$

so that $\nabla fY = df \otimes Y + f\nabla Y$.

The **torsion** of a linear connection is the $\binom{1}{2}$-tensor T^∇ on M defined by

$$T^\nabla(X, Y) := \nabla_X Y - \nabla_Y X - [X, Y],$$

and its curvature is the $\binom{1}{3}$-tensor R^∇ defined by

$$R^\nabla(X, Y)Z = \nabla_X \nabla_Y Z - \nabla_Y \nabla_X Z - \nabla_{[X,Y]} Z.$$

A linear connection gives a way to differentiate any tensor field on M along a vector field X: for a smooth function f, one defines $\nabla_X f = Xf$ and for a covariant p-tensor field α one defines

$$(\nabla_X \alpha)(Y_1, \ldots, Y_p) = X(\alpha(Y_1, \ldots, Y_p)) - \sum_{i=1}^{p} \alpha(Y_1, \ldots, \nabla_X Y_i, \ldots, Y_p).$$

Multidifferential operators on a manifold can be written in a global manner through the use of a linear connection : given a torsionfree linear connection ∇ on a manifold M (torsionfree meaning that $T^\nabla = 0$), any multidifferential operator $\mathrm{Op} : (C^\infty(M))^k \to C^\infty(M)$ writes in a unique way as

$$\mathrm{Op}(f_1, \ldots, f_k) = \sum_{J_1, \ldots, J_k} \mathrm{Op}^{J_1, \ldots, J_k} \nabla^{sym}_{J_1} f_1 \ldots \nabla^{sym}_{J_k} f_k \qquad (5)$$

where the $J_1, \ldots, J_k$ are multiindices and $\nabla^{sym}_J f$ is the symmetrised covariant derivative of order $|J|$ of f:

$$\nabla^{sym}_J f = \sum_{\sigma \in S_m} \frac{1}{m!} \nabla^m_{i_{\sigma(1)} \ldots i_{\sigma(m)}} f \qquad \text{for J} = (i_1, \ldots, i_m),$$

where $\nabla^m_{i_1 \ldots i_m} f := \nabla^m f(\partial_{i_1}, \ldots, \partial_{i_m})$ with $\nabla^m f$ defined inductively by $\nabla f := df$ and $\nabla^m f(X_1, \ldots, X_m) = (\nabla_{X_1}(\nabla^{m-1} f))(X_2, \ldots, X_m)$. The tensors $\mathrm{Op}^{J_1, \ldots, J_k}$ are covariant tensors of order $|J_1| + \ldots + |J_k|$ which are symmetric within each block of J_r indices; they are called **the tensors associated** to Op for the given connection.

Star products in the symplectic context are strongly linked to symplectic connections.

Definition 3.2. A **symplectic connection** on a symplectic manifold (M, ω) is a torsionfree linear connection ∇ which is torsionfree and such that the symplectic form ω is parallel, $\nabla \omega = 0$ (i.e. $X(\omega(Y, Z)) = \omega(\nabla_X Y, Z) + \omega(Y, \nabla_X Z))$.

It is well known that such connections exist; to see this take ∇^0 to be any torsionfree linear connection (for instance, the Levi Civita connection associated to a metric g on M), and define

$$\nabla_X Y := \nabla^0_X Y + \frac{1}{3}N(X,Y) + \frac{1}{3}N(Y,X).$$

where $\nabla^0_X \omega(Y,Z) =: \omega(N(X,Y),Z)$. Then ∇ is symplectic.

Unlike in the Riemann case, symplectic connections are not unique. Take ∇ symplectic; then $\nabla'_X Y := \nabla_X Y + S(X,Y)$ is symplectic if and only if $\omega(S(X,Y),Z)$ is totally symmetric, so the set of symplectic connections is an affine space modelled on the space of contravariant symmetric 3-tensor fields on M.

A first result concerning the link between a star product and a connection is the observation in 1978, in the seminal paper about deformation quantisation [3] by Bayen, Flato, Fronsdal, Lichnerowicz and Sternheimer that Moyal star product can be defined on any symplectic manifold (M,ω) which admits a symplectic connection ∇ with no curvature.

Lichnerowicz [13] showed that some star products on a symplectic manifold determine a unique symplecic connection; this we generalised as follows

Proposition 3.1. *[11] A star product* $* = \sum_{r\geq 0} \nu^r C_r$ *at order 2 (i.e. satisfying associativity up to terms in ν^3) on a symplectic manifold (M,ω), such that C_1 is a bidifferential operator of order 1 in each argument and C_2 of order at most 2 in each argument, determines a unique symplectic connection $\nabla = \nabla(*)$ such that*

$$C_1 = \{\,,\} + \operatorname{ad} E_1\, m \quad C_2 = \frac{1}{2}(\operatorname{ad} E_1)^2\, m + ((\operatorname{ad} E_1)\{\,,\}) + \frac{1}{2}P^2(\nabla^2\cdot,\nabla^2\cdot) + A_2 \tag{6}$$

where m is the usual multiplication of functions, where

$$(\operatorname{ad} E\, C)(u,v) = E(C(u,v)) - C(Eu,v) - C(u,Ev)$$

for any 1-diefferential operator E and any bidifferential operaotor C, where A_2 is a skewsymmetric bidifferential operator of order 1 in each argument, and where $P^2(\nabla^2\cdot,\nabla^2\cdot)$ denotes the bidifferential operator which is given by $\sum_{ij} P^{ij} P^{i'j'} \nabla^2_{ii'} u \nabla^2_{jj'} v$ in a chart.

In particular, any natural star product $* = \sum_{r\geq 0} \nu^r C_r$ *on a symplectic manifold (M,ω) determines a unique symplectic connection.*

Reciprocally, Fedosov gave in 1985 (but appearing only in the West in the nineties [9]), a recursive construction of a star product on a symplectic manifold endowed with a symplectic connection. The first proof of the existence of a star product on any symplectic manifold had been given in 1983 by De Wilde and Lecomte [7]; this was obtained by building at the same time the star product and a special derivation of it. Fedosov's construction yields a star product such that the bidifferential operators defining it are given by universal formulas in terms of the symplectic 2-form, the curvature of the connection and all its covariant derivatives.

3.1. *Fedosov's construction*

Fedosov's construction [9] gives a star product on a symplectic manifold (M, ω), when one has chosen a symplectic connection and a sequence of closed 2-forms on M. The star product is obtained by identifying the space $C^\infty(M)[[\nu]]$ with an algebra of flat sections of the so-called Weyl bundle endowed with a flat connection.

Definition 3.3. Let (V, Ω) be a symplectic vector space. We endow the space of polynomials in ν whose coefficients are polynomials on V with Moyal star product (this is the Weyl algebra $S(V^*)[\nu]$). This algebra is isomorphic to the universal enveloping algebra of the Heisenberg Lie algebra $\mathfrak{h} = V^* \oplus \mathbb{R}\nu$ with Lie bracket

$$[y^i, y^j] = (\Omega^{-1})^{ij}\nu.$$

[Indeed both associative algebras $U(\mathfrak{h})$ and $S(V^*)[\nu]$ are generated by V^* and ν and the map sending an element of $V^* \subset \mathfrak{h}$ to the corresponding element in $V^* \subset S(V^*)$ viewed as a linear function on V and mapping $\nu \in \mathfrak{h}$ on $\nu \in \mathbb{R}[\nu] \subset S(V^*)[\nu]$ has the universal property:

$$\xi *_M \xi' - \xi' *_M \xi = [\xi, \xi'] \quad \forall \xi, \xi' \in \mathfrak{h} = V^* \oplus \mathbb{R}\nu$$

so extends to a morphism of associative algebras.]

One defines a grading on $U(\mathfrak{h})$ assigning the degree 1 to each $y^i \in V^*$ and the degree 2 to ν. The **formal Weyl algebra** W is the completion in that grading of the above algebra. An element of the formal Weyl algebra is of the form

$$a(y, \nu) = \sum_{m=0}^{\infty} \left(\sum_{2k+l=m} a_{k,i_1,\dots,i_l} \nu^k y^{i_1} \dots y^{i_l} \right).$$

The product in $U(\mathfrak{h})$ is given by the Moyal star product and is extended to W:

$$(a \circ b)(y, \nu) = \left(\exp\left(\frac{\nu}{2} P^{ij} \frac{\partial}{\partial y^i} \frac{\partial}{\partial z^j} \right) a(y, \nu) b(z, \nu) \right)\Big|_{y=z}$$

with $P^{ij} = (\Omega^{-1})^{ij}$.

The symplectic group

$$Sp(V,\Omega) := \{A : V \to V \text{ linear} \mid \Omega(Au, Av) = \Omega(u, v) \,\forall u, v \in V\}$$

acts as automorphisms of $\mathfrak{h}$ by $A \cdot f = f \circ A^{-1}$ for $f \in V^*$ and $A \cdot \nu = 0$, and this action extends to both $U(\mathfrak{h})$ and W; on the latter we denote it by ρ. It respects the multiplication $\rho(A)(a \circ b) = \rho(A)(a) \circ \rho(A)(b)$. Explicitly, we have:

$$\rho(A)(\sum_{2k+l=m} a_{k,i_1,\dots,i_l}\nu^k y^{i_1}..y^{i_l}) = \sum_{2k+l=m} a_{k,i_1,\dots,i_l}\nu^k (A^{-1})^{i_1}_{j_1}..(A^{-1})^{i_l}_{j_l} y^{j_1}..y^{j_l}.$$

To an element B in the Lie algebra $sp(V,\Omega)$ we associate the quadratic element $\overline{B} = \frac{1}{2}\sum_{ijr} \Omega_{ri}B^r_j y^i y^j \in W$. The natural action $\rho_*(B)$ is given by: $\rho_*(B)y^l = \frac{-1}{\nu}[\overline{B}, y^l]$ where $[a, b] := (a \circ b) - (b \circ a)$ for any $a, b \in W$. Since both sides act as derivations this extends to all of W as

$$\rho_*(B)a = \frac{-1}{\nu}[\overline{B}, a]. \tag{7}$$

Definition 3.4. If (M, ω) is a symplectic manifold, we can form its bundle $F_{symp}(M)$ of symplectic frames. A **symplectic frame** at the point $x \in M$ is a linear symplectic isomorphism $\xi_x : (V, \Omega) \to (T_x M, \omega_x)$. The bundle $F_{symp}(M)$ is a principal $Sp(V,\Omega)$-bundle over M (the action on the right of an element $A \in Sp(V,\Omega)$ on a frame ξ_x is given by $\xi_x \circ A$).

The associated bundle $\mathcal{W} = F_{symp}(M) \times_{Sp(V,\Omega),\rho} W$ is a bundle of algebras on M called the bundle of formal Weyl algebras, or, more simply, **the Weyl bundle**. Its sections have the form of formal series

$$a(x, y, \nu) = \sum_{2k+l\geq 0} \nu^k a_{k,i_1,\dots,i_l}(x)y^{i_1} \cdots y^{i_l} \tag{8}$$

where the coefficients $a_{k,i_1,\dots,i_l}$ define (in the $i's$) symmetric covariant l–tensor fields on M. We denote by $\Gamma(\mathcal{W})$ the space of those sections.

The product of two sections taken pointwise makes $\Gamma(\mathcal{W})$ into an algebra with **multiplication**

$$(a \circ b)(x, y, \nu) = \left(\exp\left(\frac{\nu}{2}\Lambda^{ij}\frac{\partial}{\partial y^i}\frac{\partial}{\partial z^j}\right) a(x, y, \nu)b(x, z, \nu)\right)\Big|_{y=z}. \tag{9}$$

The center of this algebra coincide with $C^\infty(M)[[\nu]]$.

A symplectic connection defines a connection 1-form in the symplectic frame bundle and so a connection in all associated bundles (i.e. a covariant derivative of sections). In particular we obtain a connection in $\mathcal{W}$ which we denote by ∂; it can be viewed as a map $\partial \colon \Gamma(\mathcal{W}) \to \Gamma(\mathcal{W} \otimes \Lambda^1 T^* M)$ where sections of the bundle $\mathcal{W} \otimes \Lambda^\bullet T^* M$ are $\mathcal{W}$-valued forms on M and have locally the form

$$ a = \sum_{p \geq 0, q \geq 0} a_{pq} = \sum_{2k+p \geq 0, q \geq 0} \nu^k a_{k, i_1, \ldots, i_p, j_1, \ldots, j_q} y^{i_1} \ldots y^{i_p} \, dx^{j_1} \wedge \cdots \wedge dx^{j_q} $$

with coefficients which are again covariant tensors, symmetric in $i_1, \ldots, i_p$ and anti-symmetric in $j_1, \ldots, j_q$. [In particular $a_{00} = \sum_k \nu^k a_k$ with $a_k \in C^\infty(M)$.] Such sections can be multiplied using the product in $\mathcal{W}$ and simultaneously exterior multiplication $a \otimes \omega \circ b \otimes \omega' = (a \circ b) \otimes (\omega \wedge \omega')$ and bracketed

$$ [s, s'] = s \circ s' - (-1)^{q_1 q_2} s' \circ s $$

if $s_i \in \Gamma(\mathcal{W} \otimes \Lambda^{q_i} T^* M)$.

Let Γ^i_{kl} be the Christoffel symbols of the chosen symplectic connection ∇ and let $\overline{\Gamma} := \frac{1}{2} \sum_{ijkr} \omega_{ki} \Gamma^k_{rj} y^i y^j dx^r$; then the connection in $\mathcal{W}$ is given by

$$ \partial a = da - \frac{1}{\nu} [\overline{\Gamma}, a]. $$

For any vector field X on M, the covariant derivative ∂_X is a derivation of the algebra $\Gamma(\mathcal{W})$.

As usual, the connection ∂ in $\mathcal{W}$ extends to a covariant exterior derivative on all of $\Gamma(\mathcal{W} \otimes \Lambda^\bullet T^* M)$, also denoted by ∂, by using the Leibniz rule:

$$ \partial(a \otimes \omega) = \partial(a) \wedge \omega + a \otimes d\omega. $$

The curvature of ∂ is then given by $\partial_\circ \partial$ which is a 2-form with values in $\mathrm{End}(\mathcal{W})$. If R denotes the curvature of the symplectic connection ∇:

$$ \partial_\circ \partial a = \frac{1}{\nu} [\overline{R}, a] $$

where $\overline{R} = \frac{1}{4} \sum_{ijklr} \omega_{rl} R^l_{ijk} y^r y^k \, dx^i \wedge dx^j$.

The idea is to try to modify ∂ to have zero curvature; we look for a connection D on $\mathcal{W}$, so that D_X is a derivation af the algebra $\Gamma(\mathcal{W})$ for any vectorfield X on M, and so that D is flat in the sense that $D_\circ D = 0$. Such a connection can be written as a sum of ∂ and a $\mathrm{End}(\mathcal{W})$-valued 1-form. The latter is taken in a particular form:

$$ Da = \partial a - \delta(a) - \frac{1}{\nu} [r, a] \tag{10} $$

with $\delta(a) = \sum_k dx^k \wedge \frac{\partial a}{\partial y^k} = \frac{1}{\nu}\left[\sum_{ij} -\omega_{ij} y^i dx^j, a\right]$, so that $\delta^2 = 0$, $\partial\delta + \delta\partial = 0$ and δ is a graded derivation of $\Gamma(\mathcal{W} \otimes \Lambda^\bullet T^*M)$. Then

$$D_\circ Da = \frac{1}{\nu}\left[\overline{R} - \partial r + \delta r + \frac{1}{2\nu}[r,r], a\right]$$

and $[r,r] = 2r \circ r$. So we will have a flat connection D provided we can make the first term in the bracket be a central 2-form. Introducing

$$\delta^{-1}(a_{pq}) = \begin{cases} \frac{1}{p+q}\sum_k y^k i(\frac{\partial}{\partial x^k})a_{pq} \text{ if } p+q > 0; \\ 0 \text{ if } p+q = 0. \end{cases}$$

Then $(\delta^{-1})^2 = 0$ and $(\delta\delta^{-1} + \delta^{-1}\delta)(a) = a - a_{00}$.

Theorem 3.1. *(Fedosov [9]) The equation*

$$\delta r = -\overline{R} + \partial r - \frac{1}{\nu}r^2 + \tilde{\Omega} \tag{11}$$

for a given formal series $\tilde{\Omega} = \sum_{i\geq 1} h^i \omega_i$ of closed 2-forms ω_i on M, has a unique solution $r \in \Gamma(\mathcal{W} \otimes \Lambda^1 T^*M)$ satisfying the normalization condition $\delta^{-1}r = 0$ and such that the $\mathcal{W}$-degree of the leading term of r is at least 3. It is inductively defined by

$$r = -\delta^{-1}\overline{R} + \delta^{-1}\partial r - \frac{1}{\nu}\delta^{-1}r^2 + \delta^{-1}\tilde{\Omega}. \tag{12}$$

Since D_X acts as a derivation of the pointwise multiplication of sections, the space $\mathcal{W}_D$ of flat sections will be a subalgebra of the space of sections of $\mathcal{W}$:

$$\mathcal{W}_D = \{a \in \Gamma(\mathcal{W})|Da = 0\}.$$

The importance of this space of sections comes from the fact that there is a bijection between this space $\mathcal{W}_D$ and the space of formal power series of smooth functions on M.

Theorem 3.2. *[9] Given a flat connection D, for any $a_\circ \in C^\infty(M)[[\nu]]$ there is a unique $a \in \mathcal{W}_D$ such that $a(x,0,\nu) = a_\circ(x,\nu)$. It is defined inductively by*

$$a = \delta^{-1}\delta a + a_\circ = \delta^{-1}\left(\partial a - \frac{1}{\nu}[r,a]\right) + a_\circ. \tag{13}$$

One defines **the symbol map** $\sigma : \Gamma(\mathcal{W}) \rightarrow C^\infty(M)[[\nu]]$, by $\sigma(a) = a(x,0,\nu)$. Theorem 3.2 tells us that σ is a linear isomorphism when restricted to $\mathcal{W}_D$. So it can be used to transport the algebra structure of $\mathcal{W}_D$ to $C^\infty(M)[[\nu]]$. **Fedosov's star product** is defined by :

$$a *_F^{\nabla,\Omega} b = \sigma(\sigma^{-1}(a) \circ \sigma^{-1}(b)), \quad a,b \in C^\infty(M)[[\nu]]. \tag{14}$$

Remark that its construction depends only on the choice of a symplectic connection ∇ and the choice of a series Ω of closed 2-forms on M. If the curvature and the Ω vanish, one gets back the Moyal $*$-product.

4. Star products on Poisson manifolds

In the Poisson context, generally speaking one cannot find a "Poisson connection". Indeed, if one looks for a linear connection such that $\nabla P = 0$, then the rank of the Poisson structure must be constant. So the best we can do in general is to consider a torsionfree linear connection.

We consider star products on a manifold M which are given by universal formulas when one has chosen any Poisson structure and any connection on M. By universal, we mean the following:

Definition 4.1. [2] A **universal star product** $* = m + \sum_{r\geq 1} \nu^r C_r$ will be the association to any manifold M, any torsionfree connection ∇ on M and any Poisson tensor P on M, of a differential star product

$$*^{(M,\nabla,P)} := m + \sum_{r\geq 1} \nu^r C_r^{(M,\nabla,P)}$$

where each C_r is a universal Poisson-related bidifferential operator, i.e. so that, the tensors associated to $C_r^{(M,\nabla,P)}$ for ∇ are given by universal polynomials, involving concatenations, in P, the curvature tensor R and their covariant multiderivatives.

Kontsevich proved in 1997 the existence of a star product on any Poisson manifold as a consequence of his formality theorem. He gave an explicit formula for a formality and thus for a star product on $\mathbb{R}^d$ endowed with any Poisson structure. We shall now present this result.

4.1. *Star products on Poisson manifolds and formality*

Kontsevich in [12] showed that the set of equivalence classes of star products is the same as the set of equivalence classes of formal Poisson structure. A differential star product on M is defined by a series of bidifferential operators satisfying some identities; a formal Poisson structure on a manifold M

is completely defined by a series of bivector fields P satisfying certain properties. To describe a correspondence between these objects, one introduces the algebras they belong to.

Definition 4.2. A **differential graded Lie algebra** (briefly DGLA) is a $\mathbb{Z}$-graded vector space $\mathfrak{g} = \bigoplus_{i \in \mathbb{Z}} \mathfrak{g}^i$ endowed with
• a structure of graded Lie algebra, i.e. a graded bilinear map

$$[\,,\,]: \mathfrak{g} \otimes \mathfrak{g} \to \mathfrak{g} \text{ such that } [\,a\,,\,b\,] \subset \mathfrak{g}^{\alpha+\beta}$$

which is graded skewsymmetric ($[\,a\,,\,b\,] = -(-1)^{\alpha\beta}[\,b\,,\,a\,]$) and which satisfies the graded Jacobi identities: $[\,a\,,\,[\,b\,,\,c\,]\,] = [\,[\,a\,,\,b\,]\,,\,c\,] + (-1)^{\alpha\beta}[\,b\,,\,[\,a\,,\,c\,]\,]$ for any $a \in \mathfrak{g}^\alpha$, $b \in \mathfrak{g}^\beta$ and $c \in \mathfrak{g}^\gamma$,
• together with a differential, $d: \mathfrak{g} \to \mathfrak{g}$, i.e. a linear operator of degree 1 ($d: \mathfrak{g}^i \to \mathfrak{g}^{i+1}$) which squares to zero ($d \circ d = 0$)
• satisfying the compatibility condition (Leibniz rule)

$$d[\,a\,,\,b\,] = [\,da\,,\,b\,] + (-1)^\alpha[\,a\,,\,db\,] \qquad a \in \mathfrak{g}^\alpha, b \in \mathfrak{g}^\beta.$$

Star products and the DGLA of polydifferential operators

Let A be an associative algebra with unit on a field $\mathbb{K}$; consider the complex of multilinear maps from A to itself:

$$\mathcal{C} := \sum_{i=-1}^{\infty} \mathcal{C}^i \qquad \mathcal{C}^i := \operatorname{Hom}_{\mathbb{K}}(A^{\otimes(i+1)}, A)$$

remark that the degree $|A|$ of a $(p+1)$–linear map A is equal to p.

One extends the composition of linear operators to multilinear operators; if $A_1 \in \mathcal{C}^{m_1}$, $A_2 \in \mathcal{C}^{m_2}$, then:

$$(A_1 \circ A_2)(f_1, ..., f_{m_1+m_2+1}) :=$$

$$\sum_{j=1}^{m_1} (-1)^{(m_2)(j-1)} A_1(f_1, ..., f_{j-1}, A_2(f_j, ..., f_{j+m_2}), f_{j+m_2+1}, ..., f_{m_1+m_2+1})$$

for any $(m_1 + m_2 + 1)$- tuple of elements of A. The **Gerstenhaber bracket** is defined by

$$[A_1, A_2]_G := A_1 \circ A_2 - (-1)^{m_1 m_2} A_2 \circ A_1.$$

The differential d_D is defined by

$$d_D A = -[\mu, A] = -\mu \circ A + (-1)^{|A|} A \circ \mu$$

where μ is the usual product in the algebra A.

Proposition 4.1. *The graded Lie algebra $\mathcal{C}$ together with the differential d_D is a differential graded Lie algebra.*

Here we consider the algebra $A = C^\infty(M)$, and we deal with the subalgebra of $\mathcal{C}$ consisting of multidifferential operators $\mathcal{D}_{poly}(M) := \bigoplus \mathcal{D}^i_{poly}(M)$ with $\mathcal{D}^i_{poly}(M)$ consisting of multi differential operators acting on $i + 1$ smooth functions on M and vanishing on constants. It is easy to check that $\mathcal{D}_{poly}(M)$ is closed under the Gerstenhaber bracket and under the differential d_D, so that it is a DGLA.

Proposition 4.2. *An element $C \in \nu\mathcal{D}^1_{poly}(M)[[\nu]]$ (i.e. a series of bidifferential operator on the manifold M) yields a deformation of the usual associative pointwise product of functions μ:*

$$* = \mu + C$$

which defines a differential star product on M if and only if

$$d_D C - \frac{1}{2}[C, C]_G = 0.$$

Formal Poisson structures and the DGLA of multivector fields

A k-**multivector field** is a section of the k-th exterior power $\Lambda^k TM$ of the tangent space TM; the bracket of multivectorfields is the **Schouten-Nijenhuis bracket** which extends the usual Lie bracket of vector fields

$$[X_1 \wedge \ldots \wedge X_k, Y_1 \wedge \ldots \wedge Y_l]_S$$

$$= \sum_{r=1}^{k}\sum_{s=1}^{l}(-1)^{r+s}[X_r, X_s]X_1 \wedge \ldots \widehat{X_r} \wedge \ldots \wedge X_k \wedge Y_1 \wedge \ldots \widehat{Y_s} \wedge \ldots \wedge Y_l.$$

Since the bracket of an r- and an s-multivector fields on M is an $r + s - 1$- multivector field, we define a structure of graded Lie algebra on the space $\mathcal{T}_{poly}(M)$ of multivector fields on M by setting $\mathcal{T}^i_{poly}(M)$ the set of skewsymmetric contravariant $i + 1$-tensorfields on M (observe again a shift in the grading). We shall consider here

$$[T_1, T_2]'_S := -[T_2, T_1]_S.$$

Then $\mathcal{T}_{poly}(M)$ is turned into a differential graded Lie algebra setting the differential d_T to be identically zero.

Proposition 4.3. *An element $P \in \nu\mathcal{T}^1_{poly}(M)[[\nu]]$ (i.e. a series of bivectorfields on the manifold M) defines a formal Poisson structure on M if and only if*

$$d_T P - \frac{1}{2}[P, P]'_S = 0.$$

L_∞-algebras and L_∞-morphisms

If one could construct an isomorphism of DGLA (i.e. a linear bijection which commute with the differentials and the brackets) between the algebra $\mathcal{T}_{poly}(M)$ of multivector fields and the algebra $\mathcal{D}_{poly}(M)$ of multidifferential operators, this would give a correspondence between a formal Poisson tensor on M and a formal differential star product on M. The natural map

$$U_1 : \mathcal{T}^i_{poly}(M) \longrightarrow \mathcal{D}^i_{poly}(M)$$

which extends the usual identification between vector fields and first order differential operators, is defined by:

$$U_1(X_0\wedge\ldots\wedge X_n)(f_0,\ldots,f_n) = \frac{1}{(n+1)!} \sum_{\sigma\in S_{n+1}} \epsilon(\sigma)\, X_0(f_{\sigma(0)})\cdots X_n(f_{\sigma(n)}).$$

Unfortunately this map fails to preserve the Lie structure.

One extends the notion of morphism between two DGLA to construct a morphism whose first order approximation is this map U_1. To do this one introduces the notion of L_∞-morphism.

A toy picture of our situation (finding a correspondence between a formal Poisson tensor P on M and a formal differential star product $* = \mu + C$ on M) is the following. If C and P were elements in neighborhoods of zero V_1 and V_2 of finite dimensional vector spaces, one could consider analytic vector fields X_1 on V_1, X_2 on V_2, vanishing at zero, given by $(X_1)_C = d_D C - \frac{1}{2}[C,C]_G$, $(X_2)_P = d_T P - \frac{1}{2}[P,P]'_S$ and one would be interested in finding a correspondence between zeros of X_2 and zeros of X_1. An idea would be to construct an analytic map $\phi : V_2 \to V_1$ so that $\phi(0) = 0$ and $\phi_* X_2 = X_1$. Such a map can be viewed as an algebra morphism $\phi^* : A_1 \to A_2$ where A_i is the algebra of analytic functions on V_i vanishing at zero. The vector field X_i can be seen as a derivation of the algebra A_i. A real analytic function being determined by its Taylor expansion at zero, one can look at $C(V_i) := \sum_{n\geq 1} S^n(V_i)$ as the dual space to A_i; it is a coalgebra. One view the derivation of A_i corresponding to the vector field X_i dually as a coderivation Q_i of $C(V_i)$. One is then looking for a coalgebra morphism $F : C(V_2) \to C(V_1)$ so that $F \circ Q_2 = Q_1 \circ F$.

This is generalized to the framework of graded algebras with the notion of L_∞-morphism between L_∞-algebras.

Definition 4.3. A **graded coalgebra** on the base ring $\mathbb{K}$ is a $\mathbb{Z}$–graded vector space $C = \bigoplus_{i\in\mathbb{Z}} C^i$ with a comultiplication, i.e. a graded linear map

$$\Delta : C \to C \otimes C$$

such that $\Delta(C^i) \subset \bigoplus_{j+k=i} C^j \otimes C^k$ and such that (coassociativity):

$$(\Delta \otimes \mathrm{id})\Delta(x) = (\mathrm{id} \otimes \Delta)\Delta(x)$$

for every $x \in C$.

Additional structures that can be put on an algebra can be dualized to give a dual version on coalgebras.

Definition 4.4 (The coalgebra $C(V)$). *Let V is a graded vector space over $\mathbb{K}$, $V = \bigoplus_{i \in \mathbb{Z}} V^i$ and let $|v|$ denote the degree of $v \in V$. The tensor algebra is $T(V) = \bigoplus_{n=0}^{\infty} V^{\otimes n}$ with $V^{\otimes 0} = \mathbb{K}$. It has the two quotients: the symmetric algebra $S(V) = T(V)/ < x \otimes y - (-1)^{|x||y|} y \otimes x >$, and the exterior algebra $\Lambda(V) = T(V)/ < x \otimes y + (-1)^{|x||y|} y \otimes x >$; these spaces are naturally graded associative algebras. They can be given a structure of coalgebras with comultiplication Δ defined on a homogeneous element $v \in V$ by*

$$\Delta v := 1 \otimes v + v \otimes 1$$

and extended as algebra homomorphism.
*The **reduced symmetric space** is $C(V) := S^+(V) := \bigoplus_{n>0} S^n(V)$.*

Definition 4.5. A **coderivation** of degree d on a graded coalgebra C is a graded linear map $\delta : C^i \to C^{i+d}$ which satisfies the (co–)Leibniz identity:

$$\Delta \delta(v) = \delta v' \otimes v'' + (-1)^{d|v'|} v' \otimes \delta v''$$

if $\Delta v = \sum v' \otimes v''$. This can be rewritten with the usual Koszul sign conventions $\Delta \delta = (\delta \otimes \mathrm{id} + \mathrm{id} \otimes \delta)\Delta$.

Definition 4.6. A L_∞–**algebra** is a graded vector space V over $\mathbb{K}$ and a degree 1 coderivation Q defined on the reduced symmetric space $C(V[1])$ so that

$$Q \circ Q = 0. \tag{15}$$

[Given any graded vector space V, a new graded vector space $V[k]$ is defined by shifting the grading of the elements of V by k, i.e. $V[k] = \bigoplus_{i \in \mathbb{Z}} V[k]^i$ where $V[k]^i := V^{i+k}$.]

Definition 4.7. A L_∞–**morphism** between two L_∞–algebras, $F : (V, Q) \to (V', Q')$, is a morphism

$$F : C(V[1]) \longrightarrow C(V'[1])$$

of graded coalgebras, so that $F \circ Q = Q' \circ F$.

In the same way that any algebra morphism from $S^+(V)$ to $S^+(V')$ is determined by its restriction to V and any derivation of $S^+(V)$ is determined by its restriction to V, a coalgebra–morphism F from the coalgebra $C(V)$ to the coalgebra $C(V')$ is uniquely determined by the composition of F and the projection $\pi' : C(V') \to V'$ and any coderivation Q of $C(V)$ is determined by the composition $F \circ \pi$ where π is the projection of $C(V)$ on V.

Definition 4.8. We call **Taylor coefficients of a coalgebra-morphism** $F \colon C(V) \to C(V')$ the sequence of maps $F_n \colon S^n(V) \to V'$ and **Taylor coefficients of a coderivation** Q of $C(V)$ the sequence of maps $Q_n \colon S^n(V) \to V$.

Given V and V' two graded vector spaces, any sequence of linear maps $F_n \colon S^n(V) \to V'$ of degree zero determines a unique coalgebra morphism $F \colon C(V) \to C(V')$ for which the F_n are the Taylor coefficients. Explicitely

$$F(x_1 \ldots x_n) = \sum_{j \geq 1} \frac{1}{j!} \sum_{\{1,\ldots,n\}=I_1 \sqcup \ldots \sqcup I_j} \epsilon_x(I_1,\ldots,I_j) F_{|I_1|}(x_{I_1}) \cdots F_{|I_j|}(x_{I_j})$$

where the sum is taken over $I_1 \ldots I_j$ partition of $\{1,\ldots,n\}$ and $\epsilon_x(I_1,\ldots,I_j)$ is the signature of the effect on the odd x_i's of the unshuffle associated to the partition $(I_1,\ldots,I_j)$ of $\{1,\ldots,n\}$.

Similarly, if V is a graded vector space, any sequence $Q_n \colon S^n(V) \to V, n \geq 1$ of linear maps of degree i determines a unique coderivation Q of $C(V)$ of degree i whose Taylor coefficients are the Q_n. Explicitly

$$Q(x_1 \ldots x_n) = \sum_{\{1,\ldots,n\}=I \sqcup J} \epsilon_x(I,J)(Q_{|I|}(x_I)x_J.$$

The first conditions on the Taylor coefficients Q_n to have $Q^2 = 0$ are:

- $Q_1^2 = 0$ and Q_1 is a linear map of degree 1 on V;
- $Q_2(Q_1 x.y + (-1)^{|x|-1} x.Q_1 y) + Q_1 Q_2(x.y) = 0$;
- $0 = Q_3(Q_1 x.y.z + (-1)^{|x|-1} x.Q_1 y.z + (-1)^{|x|+|y|-2} x.y.Q_1 z) + Q_1 Q_3(x.y.z) + Q_2(Q_2(x.y).z) + (-1)^{(|y|-1)(|z|-1)} Q_2(x.z).y + (-1)^{(|x|-1)(|y|+|z|-2)} Q_2(y.z).x.$

Defining

$$dx = (-1)^{|x|} Q_1 x \qquad [x,y] := \overline{Q_2}(x \wedge y) = (-1)^{|x|(|y|-1)} Q_2(x,y), \qquad (16)$$

the above relations show that d is a differential on V, and $[\,,\,]$ is a graded skewsymmetric bilinear map from $V \times V \to V$ satisfying

$$(-1)^{|x||z|}[[x,y],z] + (-1)^{|y||x|}[[y,z],x](-1)^{|z||y|}[[z,x],y] + \text{terms in } Q_3 = 0$$

and $d[x,y] = [dx,y] + (-1)^{|x|}[x,dy]$. In particular, we get:

Lemma 4.1. *Any L_∞–algebra (V,Q) so that all the Taylor coefficients Q_n of Q vanish for $n > 2$ yields a differential graded Lie algebra and vice versa.*

The first conditions for a sequence of linear maps $F_n : S^n(V[1]) \to V'[1]$ to be the Taylor coefficients of a L_∞–morphism between two L_∞–algebras (V,Q) and (V',Q'), i.e. so that $F \circ Q = Q' \circ F$ are

- $F_1 \circ Q_1 = Q'_1 \circ F_1$ so $F_1 : V \to V'$ is a morphism of complexes from (V,d) to (V',d')
- $F_1([x,y]) - [F_1x, F_1y]' = $ expression involving F_2

so there exist L_∞–morphisms between two DGLA's which are not DGLA–morphisms.

The equations for F to be a L_∞–morphism between two DGLA's (V,Q) and $(V',Q'$ (with $Q_n = 0, Q'_n = 0 \ \forall n > 2$) are

$$Q'_1 F_n(x_1 \cdot \ldots \cdot x_n) + \frac{1}{2} \sum_{\substack{U \sqcup J = \{1,\ldots,n\} \\ I, J \neq \emptyset}} \epsilon_x(I,J) Q'_2(F_{|I|}(x_I) \cdot F_{|J|}(x_J))$$

$$= \sum_{k=1}^{n} \epsilon_x(k,1,\ldots \hat{k} \ldots, n) F_n(Q_1(x_k) \cdot x_1 \cdot \ldots \hat{x_k} \ldots \cdot x_n) \qquad (*)$$

$$+ \frac{1}{2} \sum_{k \neq l} \epsilon_x(k,l,1,\ldots \hat{kl} \ldots, n) F_{n-1}(Q_2(x_k \cdot x_l) \cdot x_1 \cdot \ldots \hat{x_k}\hat{x_l} \ldots \cdot x_n).$$

Formality, formal Poisson structures, and star products

Definition 4.9. Let $\mathfrak{m} = \nu \mathbb{R}[[\nu]]$. A $\mathfrak{m}$– **point** in a L_∞ algebra (V,Q) (over a field of characteristic zero) is an element $p \in \nu C(V)[[\nu]]$ so that $\Delta p = p \otimes p$; equivalently, it is an element $p = e^v - 1 = v + \frac{v^2}{2} + \cdots$ where v is an even element in $V[1] \otimes \mathfrak{m} = \nu V[1][[\nu]]$.

A solution of the generalized Maurer-Cartan equation is a $\mathfrak{m}$–point p where Q vanishes; equivalently, it is an odd element $v \in \nu V[[\nu]]$ so that

$$Q_1(v) + \frac{1}{2}Q_2(v \cdot v) + \cdots = 0. \qquad (17)$$

If $\mathfrak{g}$ is a DGLA, it is thus an element $v \in \mathfrak{g}$ so that $dv - \frac{1}{2}[v,v] = 0$.

The image under a L_∞ morphism of a solution of the generalised Maurer-Cartan equation is again such a solution. In particular, if one builds a L_∞ morphism F betwwen the two DGLA we consider, $F : \mathcal{T}_{poly}(M) \rightarrow \mathcal{D}_{poly}(M)$, the image under F of the point $e^\alpha - 1$ corresponding to a formal Poisson tensor,

$$\alpha \in \nu\mathcal{T}^1_{poly}(M)[[\nu]] \text{ so that } [\alpha, \alpha]_S = 0, \tag{18}$$

yields a star product on M,

$$* = \mu + \sum_n F_n(\alpha^n). \tag{19}$$

Definition 4.10. Two L_∞–algebras (V, Q) and (V', Q') are **quasi-isomorphic** if there is a L_∞–morphism F so that $F_1 : V \rightarrow V'$ induces an isomorphism in cohomology.

Kontsevich has proven that if F is a L_∞–morphism between two L_∞–algebras (V, Q) and (V', Q') so that $F_1 : V \rightarrow V'$ induces an isomorphism in cohomology, then there exists a L_∞–morphism G between (V', Q') and (V, Q) so that $G_1 : V' \rightarrow V$ is a quasi inverse for F_1 .

Definition 4.11. Kontsevich's **formality** is a quasi isomorphism between the (L_∞–algebra structure associated to the) DGLA of multidifferential operators, $\mathcal{D}_{poly}(M)$, and its cohomology, which is the DGLA of multivector fields $\mathcal{T}_{poly}(M)$.

Such a formality induces a bijective correspondence between equivalence classes of formal Poisson structures and equivalence classes of star products.

4.2. *Kontsevich's formality for $\mathbb{R}^d$*

Kontsevich gave an explicit formula for the Taylor coefficients of a formality for $\mathbb{R}^d$, i.e. the Taylor coefficients F_n of an L_∞–morphism between the two DGLA's

$$F : (\mathcal{T}_{poly}(\mathbb{R}^d), Q) \rightarrow (\mathcal{D}_{poly}(\mathbb{R}^d), Q')$$

where Q corresponds to the DGLA of $(\mathcal{T}_{poly}(\mathbb{R}^d) , [,]'_S , D_T = 0)$ and Q' corresponds to the DGLA $(\mathcal{D}_{poly}(\mathbb{R}^d) , [,]_G , d_D)$, with the first coefficient

$$F_1 = U_1 : \mathcal{T}_{poly}(\mathbb{R}^d) \rightarrow \mathcal{D}_{poly}(\mathbb{R}^d)$$

$$U_1(X_0 \wedge \ldots \wedge X_n)(f_0, \ldots, f_n) = \frac{1}{(n+1)!} \sum_{\sigma \in S_{n+1}} \epsilon(\sigma) \ X_0(f_{\sigma(0)}) \cdots X_n(f_{\sigma(n)}).$$ The formula writes

$$F_n = \sum_{m \geq 0} \sum_{\vec{\Gamma} \in G_{n,m}} \mathcal{W}_{\vec{\Gamma}} B_{\vec{\Gamma}}$$

- where $G_{n,m}$ is a set of oriented admissible graphs;
  [An admissible graph $\vec{\Gamma} \in G_{n,m}$ has n aerial vertices labelled $p_1, \ldots, p_n$, and m ground vertices labelled $q_1, \ldots, q_m$. From each aerial vertex p_i, a numer k_i of arrows are issued; each of them can end on any vertex except p_i but there can not be multiple arrows. There are no arrows issued from the ground vertices. One gives an order to the vertices:$(p_1, \ldots, p_n, q_1, \ldots, q_m)$ and one gives a compatible order to the arrows, labeling those issued from p_i with $(k_1 + \ldots + k_{i-1} + 1, \ldots, k_1 + \ldots + k_{i-1} + k_i)$. The arrows issued from p_i are named $\text{Star}(p_i) = \{\overrightarrow{p_i a_1}, \ldots, \overrightarrow{p_i a_{k_i}}\}$ with $\overrightarrow{v_{k_1 + \ldots + k_{i-1} + j}} = \overrightarrow{p_i a_j}$.]

- where $B_{\vec{\Gamma}}$ associates a m–differential operator to an n–tuple of multi-vectorfields;
  [Given a graph $\vec{\Gamma} \in G_{n,m}$ and given n multivectorfields $(\alpha_1, \ldots, \alpha_n)$ on $\mathbb{R}^d$, one defines a m– differential operator $B_{\vec{\Gamma}}(\alpha_1 \cdot \ldots \cdot \alpha_n)$; it vanishes unless α_1 is a k_1–tensor, α_2 is a k_2–tensor,..., α_n is a k_n–tensor and in that case it is given by:

$$B_{\vec{\Gamma}}(\alpha_1 \cdot \ldots \cdot \alpha_n)(f_1, \ldots, f_m) = \sum_{i_1, \ldots, i_K} D_{p_1} \alpha_1^{i_1 \ldots i_{k_1}} D_{p_2} \alpha_2^{i_{k_1+1} \ldots i_{k_1+k_2}}$$

$$\ldots D_{p_n} \alpha_n^{i_{k_1 + \ldots + k_{n-1}+1} \ldots i_K} D_{q_1} f_1 \ldots D_{q_m} f_m$$

where $K := k_1 + \cdots + k_n$ and where $D_a := \Pi_{j | \vec{v_j} = \vec{a}} \partial_{i_j}$.]

- where $\mathcal{W}_{\vec{\Gamma}}$ is the integral of a form $\omega_{\vec{\Gamma}}$ over the compactification of a configuration space $C^+_{\{p_1, \ldots, p_n\} \{q_1, \ldots, q_m\}}$.
  [Consider the upper half plane $\mathcal{H} = \{z \in \mathbb{C} | Im(z) > 0\}$; define

$$Conf^+_{\{z_1, \ldots, z_n\}\{t_1, \ldots, t_m\}} := \left\{ z_1, \ldots, z_n, t_1, \ldots, t_m \ \middle| \ \begin{array}{l} z_j \in \mathcal{H}; z_i \neq z_j \text{ for } i \neq j; \\ t_j \in \mathbb{R}; t_1 < t_2 \cdots < t_m \end{array} \right\}$$

and $C^+_{\{p_1, \ldots, p_n\}\{q_1, \ldots, q_m\}}$ to be the quotient of this space by the action of the 2-dimensional group G of all transformations of the form

$$z_j \mapsto a z_j + b \qquad t_i \mapsto a t_i + b \qquad a > 0, b \in \mathbb{R}.$$

The configuration space $C^+_{\{p_1, \ldots, p_n\}\{q_1, \ldots, q_m\}}$ has dimension $2n + m - 2$ and has an orientation induced on the quotient by

$$\Omega_{\{z_1, \ldots, z_n; t_1, \ldots, t_m\}} = dx_1 \wedge dy_1 \wedge \ldots dx_n \wedge dy_n \wedge dt_1 \wedge \ldots \wedge dt_m$$

if $z_j = x_j + i y_j$.

The compactification $\overline{C^+_{\{p_1,\ldots,p_n\}\{q_1,\ldots,q_m\}}}$ is defined as the closure of the image of the configuration space $C^+_{\{p_1,\ldots,p_n\}\{q_1,\ldots,q_m\}}$ into the product of a torus and the product of real projective spaces $P^2(\mathbb{R})$ under the map Ψ induced from a map ψ defined on $Conf^+_{\{z_1,\ldots,z_n\}\{t_1,\ldots,t_m\}}$ in the following way: to any pair of distinct points A, B taken amongst the $\{z_j, \bar{z}_j, t_k\}$ ψ associates the angle $\arg(B - A)$ and to any triple of distinct points A, B, C in that set, ψ associates the element of $P^2(\mathbb{R})$ which is the equivalence class of the triple of real numbers $(|A - B|, |B - C|, |C - A|)$.

Given a graph $\vec{\Gamma} \in G_{n,m}$, one defines a form on $\overline{C^+_{\{p_1,\ldots,p_n\}\{q_1,\ldots,q_m\}}}$ induced by

$$\omega_{\vec{\Gamma}} = \frac{1}{(2\pi)^{k_1+\ldots+k_n}(k_1)!\ldots(k_n)!} d\Phi_{\overrightarrow{v_1}} \wedge \ldots \wedge d\Phi_{\overrightarrow{v_K}}$$

where $\Phi_{\overrightarrow{p_j a}} = \mathrm{Arg}(\frac{a-p_j}{a-\bar{p}_j})$.]

We give here a sketch of the proof; a detailed proof can be found in [1].

Remark that $\mathcal{W}_{\vec{\Gamma}} \neq 0$ implies that the dimension of the configuration space $2n + m - 2$ is equal to the degree of the form $= k_1 + \ldots + k_n = K$ ($=$the number of arrows in the graph).

We shall write

$$F_n = \sum_{m \geq 0} \sum_{\vec{\Gamma} \in G_{n,m}} \mathcal{W}_{\vec{\Gamma}} B_{\vec{\Gamma}} = \sum F_{(k_1,\ldots,k_n)}$$

where $F_{(k_1,\ldots,k_n)}$ corresponds to the graphs $\vec{\Gamma} \in G_{n,m}$ with k_i arrows starting from p_i. The formality equation reads:

$$0 = F_{(k_1,\ldots,k_n)}(\alpha_1 \cdots \alpha_n) \circ \mu - (-1)^{\sum k_i - 1} \mu \circ F_{(k_1,\ldots,k_n)}(\alpha_1 \cdots \alpha_n)$$
$$+ \sum_{\substack{U \sqcup J = \{1,ldots,n\} \\ I, J \neq \emptyset}} \epsilon_\alpha(I, J)(-1)^{(|k_I|-1)|k_J|} F_{(k_I)}(\alpha_I) \circ F_{(k_J)}(\alpha_J)$$
$$- \sum_{i \neq j} F \epsilon_x(i, j, 1, ..\hat{i}\hat{j}.., n) F_{(k_i+k_j-1, k_1, .\hat{k}_i\hat{k}_j., k_n)}((\alpha_i \bullet \alpha_j) \cdot \alpha_1 \cdots .\hat{\alpha}_i\hat{\alpha}_j.. \cdot \alpha_n)$$

where

$$\alpha_1 \bullet \alpha_2 = \frac{k_1}{(k_1)!(k_2)!} \alpha_1^{r i_1 \ldots i_{k_1}-1} \partial_r \alpha_2^{j_1 \ldots j_{k_2}} \partial_{i_1} \wedge \cdots \wedge \partial_{i_{k_1}-1} \wedge \partial_{j_1} \wedge \cdots \wedge \partial_{j_{k_2}}$$

so that $[\alpha_1, \alpha_2]_S = (-1)^{k_1-1}\alpha_1 \bullet \alpha_2 - (-1)^{k_1(k_2-1)}\alpha_2 \bullet \alpha_1$. The right hand side of the formality equation can be written as

$$\sum_{\vec{\Gamma}'} C_{\vec{\Gamma}'} B_{\vec{\Gamma}'}(\alpha_1 \cdots \alpha_n)$$

for graphs $\vec{\Gamma}'$ with n aerial vertices, m ground vertices and $2n + m - 3$ arrows.

To a face G of codimension 1 in the boundary of $\overline{C^+_{\{p_1,\dots,p_n\}\{q_1,\dots,q_m\}}}$ and an oriented graph $\overrightarrow{\Gamma'}$ as above, one associates one term in the formality equation (or 0).

- $G = \partial_{\{p_{i_1},\dots,p_{i_{n_1}}\}\{q_{l+1},\dots,q_{l+m_1}\}}C^+_{\{p_1,\dots s,p_n\}\{q_1,\dots,q_m\}}$ if the aerial points $\{p_{i_1},\dots,p_{i_{n_1}}\}$ and the ground points $\{q_{l+1},\dots,q_{l+m_1}\}$ all collapse into a ground point q. We associate to G the operator $B'_{\overrightarrow{\Gamma'},G}(\alpha_1 \cdot \ \cdot \alpha_n)$ which is the term in the formality equation of the form $B_{\overrightarrow{\Gamma'}}$ obtained from

$$B_{\overrightarrow{\Gamma_2}}(\alpha_{j_1} \cdot \ \cdot \alpha_{j_{n_2}})(f_1,\dots,f_l, B_{\overrightarrow{\Gamma_1}}(\alpha_{i_1} \cdot \ \cdot \alpha_{i_{n_1}})(f_{l+1},\dots,f_{l+m_1}), f_{l+m_1+1},\dots,f_m)$$

where $\overrightarrow{\Gamma_1}$ is the restriction of $\overrightarrow{\Gamma'}$ to $\{p_{i_1},\dots,p_{i_{n_1}}\} \cup \{q_{l+1},\dots,q_{l+m_1}\}$, where $\overrightarrow{\Gamma_2}$ is obtained from $\overrightarrow{\Gamma'}$ by collapsing $\{p_{i_1},\dots,p_{i_{n_1}}\} \cup \{q_{l+1},\dots,q_{l+m_1}\}$ into q and where $\{j_1 < \dots < j_{n_2}\} = \{1,\dots,n\} \setminus \{i_1,\dots,i_{n_1}\}$.

- $G = \partial_{\{p_i,p_j\}}C^+_{\{p_1,\dots,p_n\}\{q_1,\dots,q_m\}}$ if the aerial points $\{p_i,p_j\}$ collapse into an aerial point p. If the arrow $\overrightarrow{p_ip_j}$ belongs to $\overrightarrow{\Gamma'}$, we associate to G the operator $B'_{\overrightarrow{\Gamma'},G}(\alpha_1 \cdot \ \cdot \alpha_n)$ which is the term in the formality equation of the form $B_{\overrightarrow{\Gamma'}}$ obtained from

$$B_{\overrightarrow{\Gamma_2}}(\alpha_i \bullet \alpha_j) \cdot \alpha_1 \cdot \ \hat{\alpha}_i\hat{\alpha}_j \ \cdot \alpha_n)$$

where $\overrightarrow{\Gamma_2}$ is obtained from $\overrightarrow{\Gamma'}$ by collapsing $\{p_i,p_j\}$ into p, discarding the arrow $\overrightarrow{p_ip_j}$.

If $\overrightarrow{p_ip_j}$ is not an arrow in $\overrightarrow{\Gamma'}$, we set $B'_{\overrightarrow{\Gamma'},G}(\alpha_1 \cdot \ \cdot \alpha_n) = 0$.

- $G = \partial_{\{p_{i_1},\dots,p_{i_{n_1}}\}}C^+_{\{p_1,\dots,p_n\}\{q_1,\dots,q_m\}}$ if the aerial points $\{p_{i_1},\dots,p_{i_{n_1}}\}$ all collapse with $n_1 > 2$. We associate to such a face G, the operator $B'_{\overrightarrow{\Gamma'},G} = 0$.

Looking at the coefficients of $B_{\overrightarrow{\Gamma'}}$ in each of the $B'_{\overrightarrow{\Gamma'},G}$, the right hand side of the formality equation now writes

$$\sum_{\overrightarrow{\Gamma'}} C_{\overrightarrow{\Gamma'}}B_{\overrightarrow{\Gamma'}}(\alpha_1 \cdot \ \cdot \alpha_n) = \sum_{\overrightarrow{\Gamma'}} \sum_{G \subset \partial C^+} B'_{\overrightarrow{\Gamma'},G}(\alpha_1 \cdot \ \cdot \alpha_n)$$

$$= \sum_{\overrightarrow{\Gamma'} \in G_{n,m}} \left(\sum_{G \subset \partial C^+} \int_G \omega_{\overrightarrow{\Gamma'}} \right) B_{\overrightarrow{\Gamma'}}(\alpha_1 \cdot \ \cdot \alpha_n) = 0$$

by Stokes theorem on the manifold with corners which is the compactification of C^+.

Observe that the explicit formula for the Taylor's coefficients F_j of Kontsevich L_∞ morphism from the differential graded Lie algebra of polyvectorfields on $\mathbb{R}^d$ to the differential graded Lie algebra of polydifferential operators on $\mathbb{R}^d$ shows that the coefficients of the multidifferential operator $F_j(\alpha_1, \ldots, \alpha_j)$ are given by multilinear universal expressions in the partial derivatives of the coefficients of the multivectorfields $\alpha_1, \ldots, \alpha_j$.
Hence the corresponding star product

$$f *_K^P g = f\, g + \sum_{n=1}^{\infty} \frac{\nu^n}{n!} F_n(P, \ldots, P)(f,\, g) = fg + \nu P(df, dg) + O(\nu^2) \quad (20)$$

is natural and defined by bidifferential operator whose coefficients are universal polynomials of degree n in the partial derivatives of the coefficients of the tensor P.

4.3. *Universal star product and universal formality*

Kontsevich also obtained the existence of star products on a general Poisson manifold using abstract arguments. A more direct construction of a star product on a d-dimensional Poisson manifold (M, P), using Kontsevich's formality on $\mathbb{R}^d$, was given by Cattaneo, Felder and Tomassini in [6]. Using a linear torsionfree connection ∇ on the manifold M, the construction starts with the identification of the commutative algebra $C^\infty(M)$ of smooth functions on M with the algebra of flat sections of the jet bundle $E \to M$, for the Grothendieck connection D^G. Let us recall this construction.

The jet bundle and Grothendieck connection

Let M be a d-dimensional manifold and consider the **jet bundle** $E \to M$ (the bundle of infinite jet of functions) with fibers $\mathbb{R}[[y^1, \ldots, y^d]]$ (i.e. formal power series in $y \in \mathbb{R}^d$ with real coefficients) and transition functions induced from the transition functions of the tangent bundle TM:

$$E = F(M) \times_{Gl(d,\mathbb{R})} \mathbb{R}[[y^1, \ldots, y^d]] \quad (21)$$

where $F(M)$ is the frame bundle. A section $s \in \Gamma(E)$ can be written in the form

$$s = s(x; y) = \sum_{p=0}^{\infty} s_{i_1 \ldots i_p}(x) y^{i_1} \cdots y^{i_p}$$

where the $s_{i_1 \ldots i_p}$ are components of symmetric covariant tensors on M.

The exponential map for the connection ∇ gives an identification

$$\exp_x : U \cap T_x M \to M \quad y \mapsto \exp_x(y) \tag{22}$$

at each point x, of the intersection of the tangent space $T_x M$ with a neighborhood U of the zero section of the tangent bundle TM with a neighborhood of x in M.

To a function $f \in C^\infty(M)$, one associates the section f_ϕ of the jet bundle $E \to M$ given, for any $x \in M$, by the Taylor expansion at $0 \in T_x M$ of the pullback $f \circ \exp_x$; it is given by:

$$f_\phi(x; y) = f(x) + \sum_{n>0} \frac{1}{n!} \nabla^n_{i_1 \dots i_n} f(x)\, y^{i_1} \dots y^{i_n}. \tag{23}$$

The **Grothendieck connection** D^G on E is defined by:

$$D^G_X s(x; y) := \frac{d}{dt}\Big|_{t=0} s(x(t); \exp^{-1}_{x(t)}(\exp_x(y))) \tag{24}$$

for any curve $t \to x(t) \in M$ representing $X \in T_x M$ and for any $s \in \Gamma(E)$. From the definition, it is clear that D^G is flat and that $D_G f_\phi = 0$ for any $f \in C^\infty(M)$.

Introducing $\delta = \sum_i dx^i \frac{\partial}{\partial y^i}$, and $\nabla' = \sum_i dx^i \left(\partial_{x^i} - \sum_{jk} \Gamma^k_{ij} y^j \partial_{y^k} \right)$ one can write

$$D^G = -\delta + \nabla' + A, \tag{25}$$

where A is a 1-form on M with values in the fiberwize vectorfields on E,

$$A(x; y) =: \sum_{ik} dx^i A^k_i(x; y)\, \partial_{y^k} = \sum_{ik} dx^i \left(-\frac{1}{3} \sum_{rs} R^k_{ris}(x) y^r y^s + 0(y^3) \right) \partial_{y^k}. \tag{26}$$

One extends D^G to the space $\Omega(M, E)$ of E-valued forms on M:

$$D^G = -\delta + \nabla' + A \quad \text{with } \nabla' = d - \sum_{ijk} dx^i \Gamma^k_{ij} y^j \partial_{y^k}. \tag{27}$$

Introducing (as in Fedosov"s construction) $\delta^* = \sum_j y^j i(\frac{\partial}{\partial x^j})$ on $\Omega(M, E)$, we have $(\delta^*)^2 = 0, \delta^2 = 0$ and for any $\omega \in \Omega^q(M, E_p)$, i.e. a q-form of degree p in y, we have $(\delta \delta^* + \delta^* \delta)\omega = (p+q)\omega$. Hence, defining, for any $\omega \in \Omega^q(M, E_p)$

$$\delta^{-1}\omega = \frac{1}{p+q} \delta^* \omega \qquad \text{when } p + q \neq 0$$

$$= 0 \qquad \text{when } p = q = 0$$

we see that any δ-closed q-form ω of degree p in y, when $p + q > 0$, writes uniquely as $\omega = \delta\sigma$ with $\delta^*\sigma = 0$; σ is given by $\sigma = \delta^{-1}\omega$.

One proceeds by induction on the degree in y to see that the cohomoly of D^G is concentrated in degree 0 and that any flat section of E is determined by its part of degree 0 in y. Remark that given any section s of E then $s(x; y = 0)$ determines a smooth function f on M. If $D^G s = 0$, then $s - f_\phi$ is still D^G closed. By the above, Its terms of lowest order in y must be of the form $\delta\sigma$ hence must vanish since we have a 0-form. Hence we have:

Lemma 4.2. *[5] Any section of the jet bundle $s \in \Gamma(E)$ is the Taylor expansion of the pullback of a smooth function f on M via the exponential map of the connection ∇ if and only if it is horizontal for the Grothendieck-connection D^G:*

$$s = f_\phi \text{ for a } f \in C^\infty(M) \Leftrightarrow s \in \Gamma_{hor}(E) := \{\, s' \in \Gamma(E) \,|\, D^G s' = 0 \,\}. \quad (28)$$

Furthermore, the cohomology of D^G is concentrated in degree 0.

Lemma 4.3. *[2] The 1-form A on M with values in the fiberwize vector-fields on E is given by $A(x; y) =: \sum_{ik} dx^i A_i^k(x; y)\, \partial_{y_k}$ where the A_i^k are universal polynomials given by concatenations of iterative covariant derivatives of the curvature. This 1-form A is uniquely characterized by the fact that $\delta^{-1}A = 0$. and the fact that $D^G = -\delta + \nabla' + A$ is flat.*

Universal star product

The construction of a star product on any Poisson manifold by Cattaneo, Felder and Tomassini proceeds as follows: one quantize the identification of the commutative algebra of smooth fuctions on M with the algebra of flat sections of E in the following way.

A deformed algebra stucture on $\Gamma(E)[[\nu]]$ is obtained through fiberwize quantization of the jet bundle using Kontsevich star product on $\mathbb{R}^d$.

Consider the fiberwize Poisson structure on E, P_ϕ, which is given, at a point $x \in M$, by the Taylor expansion (infinite jet) at $y = 0$ of the push-forward $(\exp_x)_*^{-1}P(\exp_x y)$.

One then considers fiberwize Kontsevich star product on $\Gamma(E)[[\nu]]$:

$$\sigma *_K^{P_\phi} \tau = \sigma\tau + \sum_{n=1}^{\infty} \frac{\nu^n}{n!} F_n(P_\phi, \ldots, P_\phi)(\sigma, \tau).$$

The operator D_X^G is not a derivation of this deformed product; one constructs a flat covariant derivative of the sections of E, D, which is a derivation of $*_K^{P_\phi}$. One defines first the derivation

$$D_X^1 = X + \sum_{j=0}^{\infty} \frac{\nu^j}{j!} F_{j+1}(\hat{X}, P_\phi, \ldots, P_\phi) \tag{29}$$

where $\hat{X} := D_X^G - X$ is a vertical vectorfield on E. The connection D^1 is not flat so one deforms it by

$$D := D^1 + [\gamma, \cdot]_{*_K^{P_\phi}}$$

so that D is flat. The 1-form γ is constructed inductively using the fact that the cohomology of D^G vanishes.

The next point is to identify series of functions on M with the algebra of flat sections of this quantized bundle of algebras to define the star product on M.

This is done by buildind a map $\rho : \Gamma(E)[[\nu]] \to \Gamma(E)[[\nu]]$ so that $\rho \circ D^G = D \circ \rho$. This map is again constructed by induction using the vanishing of the cohomology.

It results [2] from the explicit expression of the form A and the operator δ^{-1} that the star product constructed in this way is universal.

Universal formality

Similarly, Dolgushev [8] constructs a L_∞ morphism from the differential graded Lie algebra of polyvectorfields on M to the differential graded Lie algebra of polydifferential operators on M.

He defines a resolution of polydifferential operators and polyvectorfields on M using the complexes $(\Omega(M, \mathcal{D}_{poly}), D_F^{\mathcal{D}_{poly}})$ and $(\Omega(M, \mathcal{T}_{poly}), D_F^{\mathcal{T}_{poly}})$ where $\mathcal{T}_{poly}$ is the bundle of formal fiberwize polyvectorfields on E and $\mathcal{D}_{poly}$ is the bundle of formal fiberwize polydifferential operators on E. A section of $\mathcal{T}_{poly}^k$ is of the form

$$\mathcal{F}(x; y) = \sum_{n=0}^{\infty} \mathcal{F}_{i_1 \ldots i_n}^{j_1 \ldots j_{k+1}}(x) y^{i_1} \ldots y^{i_n} \frac{\partial}{\partial y^{j_1}} \wedge \ldots \wedge \frac{\partial}{\partial y^{j_{k+1}}}, \tag{30}$$

where $\mathcal{F}_{i_1 \ldots i_n}^{j_1 \ldots j_{k+1}}(x)$ are coefficients of tensors, symmetric in the covariant indices $i_1, \ldots, i_n$ and antisymmetric in the contravariant indices $j_1, \ldots, j_{k+1}$. A section of $\mathcal{D}_{poly}^k$ is of the form

$$\mathcal{O}(x; y) = \sum_{n=0}^{\infty} \mathcal{O}_{i_1 \ldots i_n}^{\alpha_1 \ldots \alpha_{k+1}}(x) y^{i_1} \ldots y^{i_n} \frac{\partial^{|\alpha_1|}}{\partial y^{\alpha_1}} \otimes \ldots \otimes \frac{\partial^{|\alpha_{k+1}|}}{\partial y^{\alpha_{k+1}}}, \tag{31}$$

where the α_l are multi-indices and $\mathcal{O}_{i_1 \ldots i_n}^{\alpha_1 \ldots \alpha_{k+1}}(x)$ are coefficients of tensors symmetric in the covariant indices $i_1, \ldots, i_n$. and symmetric in each block of α_i contravariant indices.

The spaces $\Omega(M, \mathcal{T}_{poly})$ and $\Omega(M, \mathcal{D}_{poly})$ have a formal fiberwise DGLA structure: the degree of an element in $\Omega(M, \mathcal{T}_{poly})$ (resp. $\Omega(M, \mathcal{D}_{poly})$) is defined by the sum of the degree of the exterior form and the degree of the polyvector field (resp. the polydifferential operator), the bracket on $\Omega(M, \mathcal{T}_{poly})$ is defined by $[\omega_1 \otimes \mathcal{F}_1, \omega_2 \otimes \mathcal{F}_2]_{SN} := (-1)^{k_1 q_2} \omega_1 \wedge \omega_2 \otimes [\mathcal{F}_1, \mathcal{F}_2]_{SN}$ for ω_i a q_i form and $\mathcal{F}_i$ a section in $\mathcal{T}_{poly}^{k_i}$ and similarly for $\Omega(M, \mathcal{D}_{poly})$ using the Gerstenhaber bracket. The differential on $\Omega(M, \mathcal{T}_{poly})$ is 0 and the differential on $\Omega(M, \mathcal{D}_{poly})$ is defined by $\partial := [m_{pf}, .]_G$ where m_{pf} is the fiberwize multiplication of formal power series in y of E.

The differential $D_G^{\mathcal{T}_{poly}}$ is defined on $\Omega(M, \mathcal{T}_{poly})$ by

$$D_G^{\mathcal{T}_{poly}} \mathcal{F} := \nabla^{\mathcal{T}_{poly}} \mathcal{F} - \delta^{\mathcal{T}_{poly}} \mathcal{F} + [A, \mathcal{F}]_{SN} \tag{32}$$

with $\nabla^{\mathcal{T}_{poly}} \mathcal{F} = d\mathcal{F} - \left[\sum_{ijk} dx^i \Gamma_{ij}^k y^j \partial_{y_k}, \mathcal{F} \right]_{SN}$, $\delta^{\mathcal{T}_{poly}} \mathcal{F} = \left[\sum_i dx^i \frac{\partial}{\partial y^i}, \mathcal{F} \right]_{SN}$. Similarly $D_F^{\mathcal{D}_{poly}}$ is defined on $\Omega(M, \mathcal{D}_{poly})$ by

$$D_F^{\mathcal{D}_{poly}} \mathcal{O} := \nabla^{\mathcal{D}_{poly}} \mathcal{O} - \delta^{\mathcal{D}_{poly}} \mathcal{O} + [A, \mathcal{O}]_G \tag{33}$$

with $\nabla^{\mathcal{D}_{poly}}$ and $\delta^{\mathcal{D}_{poly}}$ defined as above with the Gerstenhaber bracket.

Again the cohomology is concentrated in degree 0 and a flat section $\mathcal{F} \in \mathcal{T}_{poly}$ or $\mathcal{O} \in \mathcal{D}_{poly}$ is determined by its terms $\mathcal{F}_0$ or $\mathcal{O}_0$ of order 0 in y.

We associate to a polyvector field $F \in T_{poly}^k(M)$ a section $F_\phi \in \Gamma(\mathcal{T}_{poly})$: for a point $x \in M$ one considers the Taylor expansion (infinite jet) $F_\phi(x; y)$ at $y = 0$ of the push-forward $(\exp_x)_*^{-1} F(\exp_x y)$. Clearly this definition implies that $X_\phi(f_\phi) = (Xf)_\phi$ so that F_ϕ is uniquely determined by the fact that

$$F_\phi(f_\phi^1, \ldots, f_\phi^{k+1}) = \left(F(f^1, \ldots, f^{k+1}) \right)_\phi \qquad \forall f^j \in C^\infty(M). \tag{34}$$

Similarly we associate to a differential operator $O \in D_{poly}^k(M)$ a section $O_\phi \in \Gamma(\mathcal{D}_{poly})$ determined by the fact that

$$O_\phi(f_\phi^1, \ldots, f_\phi^{k+1}) = \left(O(f^1, \ldots, f^{k+1}) \right)_\phi \qquad \forall f^j \in C^\infty(M). \tag{35}$$

Observe that $D_G^{\mathcal{T}_{poly}} F_\phi = 0$ and similarly $D_G^{\mathcal{D}_{poly}} O_\phi = 0$, hence:

Lemma 4.4. *[2, 8] A section of $\mathcal{T}_{poly}$ is $D_F^{\mathcal{T}_{poly}}$–horizontal if and only if is a Taylor expansion of a polyvectorfield on M, i.e. iff it is of the form F_ϕ for some $F \in T_{poly}^k(M)$; a section of $\mathcal{D}_{poly}$ is $D_F^{\mathcal{D}_{poly}}$–horizontal if and only if is of the form O_ϕ for some $O \in D_{poly}^k(M)$.*

Dolgushev constructs his L_∞-morphism in two steps from the fiberwize Kontsevich formality from $\Omega(M, \mathcal{T}_{poly})$ to $\Omega(M, \mathcal{D}_{poly})$ building first a twist which depends only on the curvature and its covariant derivatives, then building a contraction using the vanishing of the D_G cohomology. Hence the Taylor coefficients of this L_∞-morphism, which are. a collection of maps F_j^D associating to j multivectorfields α_k on M a multidifferential operator $F_j^D(\alpha_1, \ldots, \alpha_j)$ are such that the tensors defining this operator are given by universal polynomials in the tensors defining the α_j's, the curvature tensor and their iterated cavariant derivatives.

References

1. D. Arnal, D. Manchon et M. Masmoudi, Choix des signes pour la formalité de M. Kontsevich, *Pacific Journal of math.* 203, 1 (2002) 23–66.
2. Mourad Ammar, Véronique Chloup et Simone Gutt, Universal star products, *Lett. in Math. Phys.* 84 (2008) 199–215.
3. F. Bayen, M. Flato, C. Fronsdal, A. Lichnerowicz and D. Sternheimer, Quantum mechanics as a deformation of classical mechanics, *Lett. Math. Phys.* 1 (1977) 521–530 and Deformation theory and quantization, part I, *Ann. of Phys.* 111 (1978) 61–110.
4. F. Bayen, M. Flato, C. Fronsdal, A. Lichnerowicz and D. Sternheimer, Deformation theory and quantization, part II, *Ann. of Phys.* 111 (1978) 111–151
5. A.S. Cattaneo, G.Felder, On the globalization of Kontsevich's star product and the perturbative Poisson sigma model, Progress of Theoretical Physics Supplement No.144 (2002) 38–53.
6. A. Cattaneo, G. Felder and L. Tomassini, From local to global deformation quantization of Poisson manifolds, *Duke Math. J.* 115 (2002) 329–352.
7. M. De Wilde and P. Lecomte, Existence of star-products and of formal deformations of the Poisson Lie algebra of arbitrary symplectic manifolds, *Lett. Math. Phys.* 7 (1983) 487–496.
8. V. Dolgushev, Covariant and equivariant formality theorems, *Adv. Math.* 191 (2005) 147–177.
9. B.V. Fedosov, A simple geometrical construction of deformation quantization, *J. Diff. Geom.* 40 (1994) 213–238.
10. M. Flato, A. Lichnerowicz and D. Sternheimer, Crochet de Moyal–Vey et quantification, *C. R. Acad. Sci. Paris I Math.* 283 (1976) 19–24.

11. S. Gutt and J. Rawnsley, Natural star products on symplectic manifolds and quantum moment maps, *Lett. in Math . Phys.*66 (2003), no. 1-2, 123–139.

12. M. Kontsevich, Deformation quantisation of Poisson manifolds, I. IHES preprint q-alg/9709040,*Lett. Math. Phys.* 66 (2003) 157–216.

13. A. Lichnerowicz, Déformations d'algèbres associées à une variété symplectique (les $*_\nu$-produits), *Ann. Inst. Fourier, Grenoble* 32 (1982) 157–209.

WHAT IS SYMPLECTIC GEOMETRY?

DUSA MCDUFF

Department of Mathematics, Barnard College, Columbia University
dusa@math.columbia.edu

In this article we explain the elements of symplectic geometry, and sketch the proof of one of its foundational results — Gromov's nonsqueezing theorem — using J-holomorphic curves. The work presented here was partially supported by NSF grant DMS 0604769.

1. First notions

Symplectic geometry is an even dimensional geometry. It lives on even dimensional spaces, and measures the sizes of 2-dimensional objects rather than the 1-dimensional lengths and angles that are familiar from Euclidean and Riemannian geometry. It is naturally associated with the field of complex rather than real numbers. However, it is not as rigid as complex geometry: one of its most intriguing aspects is its curious mixture of rigidity (structure) and flabbiness (lack of structure). In this talk I will try to describe some of the new kinds of structure that emerge.

First of all, what is a symplectic structure? The concept arose in the study of classical mechanical systems, such as a planet orbiting the sun, an oscillating pendulum or a falling apple. The trajectory of such a system is determined if one knows its position and velocity (speed and direction of motion) at any one time. Thus for an object of unit mass moving in a given straight line one needs two pieces of information, the position q and velocity (or more correctly momentum) $p := \dot{q}$. This pair of real numbers $(x_1, x_2) := (p, q)$ gives a point in the plane $\mathbb{R}^2$. In this case the symplectic structure ω is an area form (written $dp \wedge dq$) in the plane. Thus it measures the area of each open region S in the plane, where we think of this region as oriented, i.e. we choose a direction in which to traverse its boundary ∂S. This means that the area is signed, i.e. as in Figure 1.1 it can be positive or negative depending on the orientation. By Stokes' theorem, this is equivalent to measuring the integral of the action $p \, dq$ round the boundary ∂S.

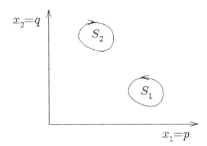

Fig. 1.1.   The area of the region S_1 is positive, while that of S_2 is negative.

This might seem a rather arbitrary measurement. However, mathematicians in the nineteenth century proved that it is preserved under time evolution. In other words, if a set of particles have positions and velocities in the region S_1 at the time t_1 then at any later time t_2 their positions and velocities will form a region S_2 with the same area. Area also has an interpretation in modern particle (i.e. quantum) physics. Heisenberg's Uncertainty Principle says that we can no longer know both position and velocity to an arbitrary degree of accuracy. Thus we should not think of a particle as occupying a single point of the plane, but rather lying in a region of the plane. The Bohr-Sommerfeld quantization principle says that the area of this region is quantized, i.e. it has to be an integral multiple of a number called Planck's constant. Thus one can think of the symplectic area as a measure of the entanglement of position and velocity.

An object moving in the plane has two position coordinates q_1, q_2 and correspondingly two velocity coordinates $p_1 = \dot{q}_1, p_2 = \dot{q}_2$ that measure its speed in each direction. So it is described by a point

$$(x_1, x_2, x_3, x_4) = (p_1, q_1, p_2, q_2) \in \mathbb{R}^4$$

in the 4-dimensional space $\mathbb{R}^4$. The symplectic form ω now measures the (signed) area of 2-dimensional surfaces S in $\mathbb{R}^4$ by adding the areas of the projections of S to the (x_1, x_2)-plane and the (x_3, x_4)-plane. Thus, as is illustrated in Figure 1.2,

$$\omega(S) = \text{area}(pr_{12}(S)) + \text{area}(pr_{34}(S)).$$

Notice that $\omega(S)$ can be zero: for example S might be a little rectangle in the x_1, x_3 directions which projects to a line under both pr_{12} and pr_{34}.

More technically, ω is a differential 2-form written as

$$\omega = dx_1 \wedge dx_2 + dx_3 \wedge dx_4,$$

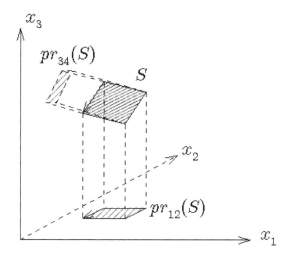

Fig. 1.2. The symplectic area $\omega(S)$ is the sum of the area of its projection $pr_{12}(S)$ to the plane $(x_1, x_2) = (p_1, q_1)$ given by the velocity and position in the first direction together with the area of the corresponding projection $pr_{34}(S)$ for the two coordinates in the second direction. I have drawn the first 3 coordinates; the fourth is left to your imagination.

and we evaluate the area $\omega(S) = \int_S \omega$ by integrating this form over the surface S. A similar definition is made for particles moving in n-dimensions. The symplectic area form ω is again the sum of contributions from each of the n pairs of directions:

$$\omega_0 = dx_1 \wedge dx_2 + dx_3 \wedge dx_4 + \cdots + dx_{2n-1} \wedge dx_{2n}. \tag{1}$$

We call this form ω_0 because it is the standard symplectic form on Euclidean space. The letter ω is used to designate any symplectic form.

To be even more technical, one can define a symplectic form ω on any even dimensional smooth (i.e. infinitely differentiable) manifold M as a closed, nondegenerate 2-form, where the nondegeneracy condition is that for each nonzero tangent direction v there is another direction w such that the area $\omega(v, w)$ of the little (infinitesimal) parallelogram spanned by these vectors is nonzero. (For a geometric interpretation of these conditions on ω see Figure 1.3.) A manifold is said to *be symplectic* or to *have a symplectic structure* if it is provided with a symplectic form.

The first important theorem in symplectic geometry is that locally[a] all symplectic forms are the same.

---

[a]This means "on suitably small open sets".

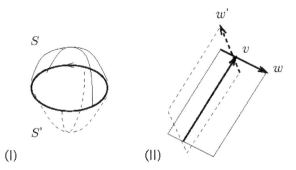

(I)   (II)

Fig. 1.3. The fact that ω is *closed* means that the symplectic area of a surface S with boundary does not change as S moves, provided that the boundary is fixed. Thus in (I) the surfaces S and S' have the same area. Diagram (II) illustrates the *nondegeneracy* condition: for any direction v at least one of the family of 2-planes spanned by v and a varying other direction w has non zero area.

Darboux's Theorem: *Given a symplectic form ω on a manifold M and any point on M one can always find coordinates $(x_1, \ldots, x_{2n})$ defined in an open neighborhood U of this point such that in this coordinate system ω is given on the whole open set U by formula (1).*

This is very different from the situation in the usual (Riemannian) geometry where one can make many local measurements (for example involving curvature) that distinguish among different structures. Darboux's theorem says that *all* symplectic structures are locally indistinguishable. Of course, as mathematicians have been discovering in the past 20 years, there are many very interesting global invariants that distinguish different symplectic structures. But most of these are quite difficult to define, often involving deep analytic concepts such as the Seiberg–Witten equations or J-holomorphic curves.

A symplectic form ω has an important invariant, called its cohomology class $[\omega]$. This class is determined by the areas $\omega(S)$ of *all* closed[b] surfaces S in M. In fact, for compact M the class $[\omega]$ is determined by a finite number of these areas $\omega(S_i)$ and so contains only a finite amount of information. Cf. Figure 1.3 where we pointed out that the area $\omega(S)$ does not change if we move S around.

There is a similar flabbiness in the symplectic structure itself. A fundamental theorem due to Moser says that one cannot change the symplectic

---

[b]A *closed* surface is something like the surface of a sphere or donut; it has no edges and no holes.

form in any important way by deforming it, provided that the cohomology class is unchanged. More precisely, if $\omega_t, t \in [0,1]$, is a smooth path of symplectic forms such that $[\omega_0] = [\omega_t]$ for all t, then all these forms are "the same" in the sense that one can make them coincide by moving the points of M appropriately.[c] The important point here is that we cannot find new structures by deforming the old ones, provided that we fix the integrals of our forms over all closed surfaces. This result is known as **Moser's Stability Theorem**, and is an indication of robustness of structure.[d]

2. Symplectomorphisms

Another consequence of the lack of local features that distinguish between different symplectic structures is that there are many ways to move the points of the underlying space M without changing the symplectic structure ω. Such a movement is called a *symplectomorphism*. This means first that

• ϕ is a *diffeomorphism*, that is, it is a bijective (one to one and onto) and smooth (infinitely differentiable) map $\phi : M \to M$, giving rise to the movement $x \mapsto \phi(x)$ of the points x of the space M;

and second that

• it *preserves symplectic area*, i.e. $\omega(S) = \omega(\phi(S))$ for *all* little pieces of surface S. The important point here is that this holds for all S, no matter how small or large. (Technically it is better to work on the infinitesimal level, looking at the properties of the derivative $d\phi$ of ϕ at each point.)

In 2-dimensions, a symplectomorphism ϕ is simply an area preserving transformation. For example the map ψ in Figure 2.1 is given by the formula $\psi(x_1, x_2) = (2x_1, \frac{1}{2}x_2)$. Since it multiplies one coordinate by two and divides the other by two it does not change area. More generally, one form of Moser's theorem says the following:

Characterization of plane symplectomorphisms: *Suppose that S is a region in the plane $\mathbb{R}^2$ that is diffeomorphic to a disc D and has the same area as D. Then there is a symplectomorphism $\phi : D \to S$.*

The above statement means that we can choose the diffeomorphism $\phi : D \to S$ so that it preserves the area of *every* subset of D not just of D itself.

---

[c]In technical language we say that these forms are all diffeomorphic.

[d]For precise statements, many proofs, and a list of references on all the topics mentioned here see [9]. There are also other more elementary books such as Cannas [2]. For simplicity, we shall only work here in dimensions 2 and 4. But all the results have higher dimensional analogs.

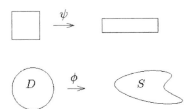

Fig. 2.1.   Symplectomorphisms in dimension 2.

In 4-dimensions the situation is rather different. Gromov was the first to try to answer the question:

What are the possible shapes of a symplectic ball?

More precisely, let B be the round ball of radius one in $\mathbb{R}^4$. Thus

$$B = \left\{(x_1,\ldots,x_4) : x_1^2 + x_2^2 + x_3^2 + x_4^2 \leq 1\right\}$$

consists of all points whose (Euclidean) distance from the origin $\{0\}$ is at most one. What can one say about the set $\phi(B)$ where ϕ is any symplectomorphism? Can $\phi(B)$ be long and thin? Can its shape be completely arbitrary? The analog of the 2-dimensional result would be that $\phi(B)$ could be any set that is diffeomorphic to B and also has the same volume.[e]

It is possible for $\phi(B)$ to be long and thin. For example one can stretch out the coordinates x_1, x_3 while shrinking the pair x_2, x_4 as in the map

$$\phi\big((x_1, x_2, x_3, x_4)\big) = (2x_1, \tfrac{1}{2}x_2, 2x_3, \tfrac{1}{2}x_4).$$

But the map

$$\psi\big((x_1, x_2, x_3, x_4)\big) = (\tfrac{1}{2}x_1, \tfrac{1}{2}x_2, 2x_3, 2x_4)$$

will *not* do since the area of rectangles in the x_1, x_2 plane are divided by 4. Note that ϕ preserves the pairs (x_1, x_2) and (x_3, x_4) and is made by combining area preserving transformations in each of these 2-planes. One might ask if there is a symplectomorphism that mixes these pairs, for example rotates in the x_1, x_3 direction. There certainly are such maps. For example

$$\phi\big((x_1, x_2, x_3, x_4)\big) = \tfrac{1}{2}(x_1 - x_3,\ x_2 - x_4,\ x_1 + x_3,\ x_2 + x_4)$$

---

[e]Since $\omega \wedge \omega$ is a volume form, any symplectomorphism preserves volume. The fact that it is impossible to give a completely elementary proof of this (e.g. one that does not involve the concept of a differential form) reflects the fact that to nineteenth century mathematicians this was a nontrivial result; cf. [1].

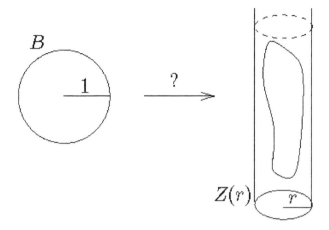

Fig. 2.2. Can the unit ball B be squeezed into the cylinder $Z(r)$?

is a symplectomorphism that rotates anticlockwise by 45 degrees in both the x_1, x_3 plane and the x_2, x_4 plane.

Nevertheless, Gromov in his nonsqueezing theorem showed that as far as the large geometric features of the space are concerned one can still see this splitting of $\mathbb{R}^4$ into the product of the (x_1, x_2) plane and the (x_3, x_4) plane. He described this in terms of maps of the unit ball B into the cylinder

$$Z(r) := D^2(r) \times \mathbb{R}^2 = \left\{(x_1, \ldots, x_4) : x_1^2 + x_2^2 \leq r^2\right\} \subset \mathbb{R}^4 \qquad (1)$$

of radius r, showing that one cannot squeeze a large ball into a thin cylinder of this form.

Gromov's Nonsqueezing Theorem: *If $r < 1$ there is no symplectomorphism ϕ such that $\phi(B) \subset Z(r)$.*

Although the nonsqueezing theorem might seem quite special and therefore unimportant (though perhaps cute), the property expressed here, that symplectomorphisms cannot squeeze a set in a pair of "symplectic directions" such as x_1, x_2, turns out to be absolutely fundamental: when properly formulated it gives a necessary and sufficient condition for a diffeomorphism to preserve the symplectic structure. Thus this theorem should be understood as a geometric manifestation of the very nature of a symplectic structure.

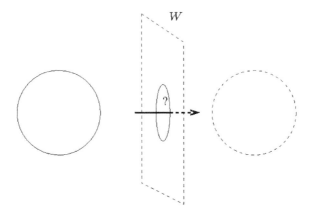

Fig. 2.3.   Can the ball go through the hole?

Another similar problem is that of the *Symplectic Camel*.[f] Here the camel is represented by a round 4-dimensional ball of radius 1 say, and the eye of the needle is represented by a "hole in a wall". That is to say, the wall with a hole removed is given by

$$W = \left\{ (x_1, x_2, x_3, x_4) \in \mathbb{R}^4 \ : \ x_1 = 0, x_2^2 + x_3^2 + x_4^2 \geq 1 \right\}$$

and we ask whether a (closed) round ball B of radius 1 can be moved from one side of the wall to the other in such a way as to preserve the symplectic form. (Note that because the ball is closed and the 2-sphere $\{x_2^2 + x_3^2 + x_4^2 = 1, x_1 = 0\}$ is contained in the wall, the ball will get stuck half way if one just tries moving it by a translation.)

It is possible to do this if one just wants to preserve volume. This is easy to see if one restricts to the three-dimensional case (by forgetting the last coordinate x_4); one can imagine squeezing a sufficiently flexible balloon through any small hole while preserving its volume. However, as Gromov showed, the symplectic case is more rigid.

The Symplectic Camel: *It is impossible to move a ball of radius ≥ 1 symplectically from one side of the wall to the other.*

Both these results show that there is some rigidity in symplectic geometry. Exactly how this is expressed is still not fully understood, especially

---

[f]The name of this problem is a somewhat "in" joke, of the kind appreciated by many mathematicians. The reference is to the saying that it would be easier for a camel to go through the eye of needle than for a rich man to get into heaven. (This saying is probably a mistranslation of a sentence in the bible.)

in dimensions > 4. However, recently progress has been made on another fundamental embedding problem in dimension 4. The question here is to understand the conditions under which one ellipsoid embeds symplectically[g] in another. Here, by the ellipsoid $E(a, b)$ we mean the set

$$E(a,b) := \left\{ (x_1, \ldots, x_4) \; : \; \frac{x_1^2 + x_2^2}{a} + \frac{x_3^2 + x_4^2}{b} \le 1 \right\}.$$

In this language, the ball $B(r)$ of radius r is $E(r^2, r^2)$; in other words, the numbers a, b are proportional to areas, not lengths. Thus the question is: when does $E(a, b)$ embed symplectically in $E(a', b')$? Here we will fix notation by assuming that $a \le b$ and $a' \le b'$.

If the first ellipsoid is a ball $E(a, a)$ then the answer is given by the Nonsqueezing Theorem:

> *a necessary and sufficient condition for embedding* $E(a, a)$ *into* $E(a', b')$ *(where* $a' \le b'$*) is that* $a \le a'$.

(This condition is obviously sufficient since $E(a, b)$ is a subset of $E(a', b')$ when $a \le a'$ and $b \le b'$. On the other hand, it is necessary because if $E(a, a)$ embeds in $E(a', b')$ then, since $E(a', b') \subset Z(\sqrt{a'})$, it also embeds in $Z(\sqrt{a'})$, so that by the nonsqueezing theorem we must have $a \le a'$.)

But if the target is a ball $E(a', a')$ and the domain $E(a, b)$ is an arbitrary ellipsoid the answer is not so easy. It was proved in the 90s that when $a \le b \le 2a$ the situation is rigid: to embed $E(a, b)$ into $E(a', a')$ it is necessary and sufficient that $b \le a'$. In other words, the ellipsoid does *not* bend in this case. However, as soon as $b > 2a$ some flexibility appears and it is possible to embed $E(a, b)$ into $E(a', a')$ for some $a' < b$. Then of course one wants to know how much flexibility there is. What other obstructions are there to performing such an embedding besides the obvious one of volume? Notice that because a, b, a' are areas the volume obstruction to the existence of an embedding is that $ab \le (a')^2$.

This question was nicely formulated in a paper by Cieliebak, Hofer, Latschev and Schlenk [3] called *Quantitative Symplectic Geometry* in terms of the following function: define $c(a)$ for $a \ge 1$ by[h]

$$c(a) := \inf\{c' : E(a, b) \text{ embeds symplectically in } E(c', c')\}.$$

---

[g]We say that the set U embeds symplectically in V if there is a symplectomorphism ϕ such that $\phi(U) \subset V$.

[h]Note that $E(a, b)$ may not embed in $E(c(a), c(a))$ itself – one usually needs a little extra room so that the boundary of $E(a, b)$ does not fold up on itself. However, one can show that the *interior* of $E(a, b)$ does embed in $E(c(a), c(a))$.

E(16,25)   E(20,20)

Fig. 2.4.   Does $E(a,b)$ embed symplectically in $E(a',b')$? In the case illustrated here, $ab = a'b'$, so there is no volume obstruction to the embedding, but the embedding does not exist because $25 > 20$. In fact, if we rescale by dividing all areas by 16, the problem is equivalent to embedding $E(1, \frac{25}{16})$ into $E(\frac{20}{16}, \frac{20}{16})$. But this is impossible by Equation (2) below.

When [3] was written, this function was largely a mystery except that one knew that $c(a) = a$ for $a \leq 2$ (rigidity). Now methods have been developed to understand it, and it should be fully known soon for all a: see McDuff and Schlenk [12] and also [11]. As a first step, work of Opshtein [14] can be used to evaluate $c(a)$ in the range $1 \leq a \leq 4$. Surprisingly, it turns out that

$$c(a) = a, \quad \text{if } 1 \leq a \leq 2, \quad c(a) = 2 \text{ if } 2 \leq a \leq 4. \tag{2}$$

In other words the graph is *constant* in the range $a \in [2,4]$. To prove this one only needs to show that $c(4) = 2$. Because c is nondecreasing, if $c(2) = c(4) = 2$ then c must be constant on this interval. On the other hand, the statement $c(4) = 2$ implies that we can fill the volume of the ball $E(2,2)$ by the interior of the ellipsoid $E(1,4)$, which is a somewhat paradoxical state of affairs. Why can you fill all the volume of $E(2,2)$ by the interior of $E(1,4)$ when you cannot fill $E(\sqrt{2}, \sqrt{2})$ by the interior of $E(1,2)$?

It turns out to be important that 4 is a perfect square. For any positive integer k, Opshtein discovered an explicit way to embed $E(1, k^2)$ into $E(k, k)$ that embeds the ellipsoid $E(1, k^2)$ into a neighborhood of a degree k curve such as $z_0^k + z_1^k + z_2^k = 0$ in the complex projective plane. Thus there is a clear geometric reason why the case $a = k^2$ is different from the general case. Many more things are now known about the function c: Figure 2.5 gives an idea of its graph. Here I would just like to point out that its behavior on the interval $[1,4]$ is typical in symplectic geometry: either the situation is rigid (for $a \in [1,2]$, the ellipse does not bend at all) or it is as flexible as it could possibly be (for $a \in [2,4]$, the ellipsoid bends as much as is consistent with the obvious constraints coming from volume and the constraint at $a = 2$).

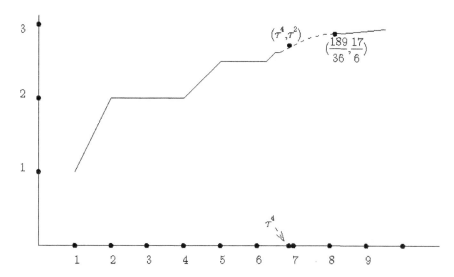

Fig. 2.5. The graph of c: it appears to have an infinite staircase that converges to the point (τ^4, τ^2), where τ is the golden ratio, and it equals $\sqrt{a}$ for $a \geq \frac{189}{36}$. The graph between the points $\tau^4 < a < \frac{1}{36}$ is not yet completely known.

3. Almost complex structures and J-holomorphic curves

The remainder of this note tries to give a rough idea of how Gromov proved his results. Nowadays there are many possible approaches to the proof. But we shall explain Gromov's original idea that uses J-holomorphic curves. These provide a special way of cutting the cylinder into 2-dimensional slices of area πr^2 as in Figure 3.5, and we shall see that these provide an obstruction to embedding a ball of radius 1. Similarly, because one can fill the hole in the wall by these slices, the size of a ball that can be moved through the hole is constrained to be < 1.

The concept of J-holomorphic curves has turned out to be enormously fruitful. Gromov's introduction of this idea in 1985 was one of the main events that initiated the modern study of symplectic geometry.

Gromov's key idea was to exploit the connection between symplectic geometry and the complex numbers.

A differentiable manifold M is a space in which one can do calculus: locally it looks like Euclidean space, but it can have interesting global structure.[i] As in calculus, one often approximates curves or surfaces near a given

---

[i]For a wonderful introduction to this subject see Milnor [13].

point $x \in M$ by the closest linear objects, tangent lines or planes as the case may be. The collection of all possible tangent directions at a point x is called the tangent space $T_x M$ to M at x. It is a linear (or vector) space of the same dimension as M. As the point x varies over M the collection $\cup_{x \in M} T_x M$ of all these planes forms what is called the *tangent bundle* of M. If $M = \mathbb{R}^{2n}$ is Euclidean space itself, then one can identify each of its tangent spaces $T_x \mathbb{R}^{2n}$ with $\mathbb{R}^{2n}$, but most manifolds (such as the sphere) curve around and do not contain their tangent spaces.

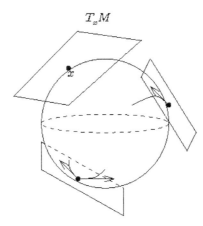

Fig. 3.1.   Some curves and tangent vectors on the two-sphere, together with some tangent spaces $T_x M$.

An *almost complex structure at a point x* of a manifold M is a linear transformation J_x of the tangent space $T_x M$ at x whose square is -1. Geometrically, J_x rotates by a quarter turn (with respect to a suitable coordinate system at x.) Thus the tangent space $T_x M$ becomes a complex vector space (with the action of J_x playing the role of multiplication by $\sqrt{-1}$.) An *almost complex structure J* on M is a collection J_x of such transformations, one for each point of x, that varies smoothly as a function of x. If M has dimension 2 one can always choose local coordinates on M to make the function $x \to J_x$ constant. However in higher dimensions this is usually impossible. If such coordinates exist J is said to be *integrable*. What this means is explained more fully in Equation (2).

Rather few manifolds have integrable almost complex structures. (To be technical for a minute, this happens if and only if M has a complex structure, i.e. if and only if one can glue M together from its locally Euclidean

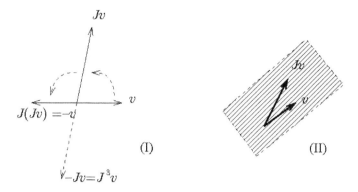

Fig. 3.2. (I) pictures $J = J_x$ as a (skew) rotation; (II) shows the complex line spanned by v, Jv.

pieces $U \subset \mathbb{C}^n$ by using holomorphic functions.) However, many manifolds have almost complex structures.[j]

In particular, symplectic manifolds always do. In fact, in this case one can choose J to be *compatible* with the symplectic form ω, i.e. so that at all points $x \in M$

$$\omega(J_x v, J_x w) = \omega(v, w), \quad \text{and} \quad \omega(v, J_x v) > 0, \tag{1}$$

for all nonzero tangent vectors $v, w \in T_x M$.[k]

The first equation here says that rotation by J_x preserves symplectic area, while the inequality (called the *taming condition*) says that every complex line has positive symplectic area. Note that complex lines have 2 real dimensions; they are spanned over $\mathbb{R}$ by two vectors of the form v and $J_x v = $ "iv".

For any given ω there are many compatible almost complex structures; in fact there is a contractible set of such structures. Associated to each such J there is a Riemannian metric, i.e. a symmetric inner product g_J on the tangent space $T_x M$. It is given by the formula

$$g_J(v, w) := \omega(v, Jw), \quad v, w \in T_x M.$$

As with any metric, this gives a way of measuring lengths and angles. However, it depends on the choice of J and so is not determined by ω

---

[j]For example the 6-dimensional sphere S^6 has an almost complex structure. It is a famous unsolved problem to decide whether it has a complex structure.

[k]Here I have used the language of differential 2-forms; but readers can think of $\omega(v, w)$ as the symplectic area of a small (infinitesimal) parallelogram spanned by the vectors v, w.

alone. Nevertheless, because via J it has a very geometric relationship to ω it is often useful to consider it.[1]

As an example that will be useful later, observe that the usual (integrable!) complex structure J_0 on $\mathbb{C}^2 = \mathbb{R}^4$ is compatible with the standard symplectic form ω_0 and the associated metric g_0 is the usual Euclidean distance function.

J-holomorphic curves: A (real) curve in a manifold M is a path in M; that is, it is the image of a map $f : U \to M$ where U is a subinterval of the real line $\mathbb{R}$. A J-holomorphic curve in an almost complex manifold (M, J) is the complex analog of this. It has one complex dimension (but 2 real dimensions) and is the image of a "complex" map $f : \Sigma \to M$ from some complex curve Σ into (M, J). Here we shall take the domain Σ to be either a 2-dimensional disc D (consisting of a circle in the plane together with its interior) or the 2-sphere $S^2 = \mathbb{C} \cup \{\infty\}$, which we shall think of as the complex plane $\mathbb{C}$ completed by adding a point at ∞; see Figure 3.3.

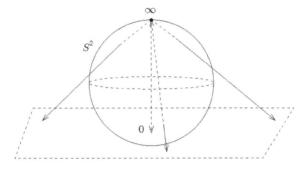

Fig. 3.3.   The 2-sphere S^2 as the completion of the complex plane $\mathbb{C}$. Often one puts ∞ at the north pole of the 2-sphere and identifies $S^2 \setminus \{\infty\}$ with the plane via stereographic projection.

If J is integrable, we can choose local complex coordinates on the target space M of the form $z_1 = x_1 + ix_2$, $z_2 = x_3 + ix_4$ so that at each point x the linear transformation $J = J_x$ acts on the tangent vectors $\frac{\partial}{\partial x_j}$ by "multiplication by i": namely,

$$J\left(\frac{\partial}{\partial x_1}\right) = \frac{\partial}{\partial x_2}, \quad J\left(\frac{\partial}{\partial x_2}\right) = -\frac{\partial}{\partial x_1}, \quad J\left(\frac{\partial}{\partial x_3}\right) = \frac{\partial}{\partial x_4}, \quad J\left(\frac{\partial}{\partial x_4}\right) = -\frac{\partial}{\partial x_3}. \quad (2)$$

---

[1]For example, the associated metric on the loop space of M leads to a very natural interpretation for gradient flows on this loop space. This is the basis of Floer theory; see [8].

Then there is an obvious notion of "complex" map: in terms of a local coordinate $z = x + iy$ on the domain and this coordinate system on the target, f is given by two power series $f_1(z)$, $f_2(z)$ with complex coefficients a_k, b_k:

$$f(z) = \big(f_1(z), f_2(z)\big) = \Big(\sum_{k \geq 0} a_k z^k, \sum_{k \geq 0} b_k z^k \Big),$$

i.e. f is holomorphic. Such functions can be characterized by the behavior of their derivatives: these must satisfy the Cauchy–Riemann equation

$$\frac{\partial f}{\partial x} + J \frac{\partial f}{\partial y} = 0.$$

This equation still makes sense even if J is not integrable, and so given such J we say that $f : \Sigma \to (M, J)$ is J-holomorphic if it satisfies the above equations.

J-holomorphic curves as minimal surfaces: The images $f(\Sigma)$ of such maps have very nice properties. In particular, their area with respect to the associated metric g_J equals their symplectic area. We saw earlier that the symplectic area of a surface is invariant under deformations of the surface that fix its boundary. (Cf. Figure 1.3 (I).) It follows easily that their metric area can only *increase* under such deformations, i.e. J-holomorphic curves are so-called g_J-minimal surfaces. Thus we can think of them as the complex analog of a real geodesic.[m]

Minimal surfaces have the following very important property that we will use later. Let g_0 be the usual Euclidean metric on $\mathbb{R}^4$ (or, in fact, on any Euclidean space $\mathbb{R}^d$). Suppose that S is a g_0-minimal surface in the ball B of radius 1 that goes through the center of the ball and has the property that its boundary lies on the surface of the ball. (Technically, we say that S is *properly embedded* in B.) Then

$$\text{the } g_0\text{-area of } S \text{ is } \geq \pi. \tag{3}$$

In fact the g_0-minimal surface of *least* area that goes through the center of a unit ball is a flat disc of area π. All others have nonpositive curvature, which means that at each point they bend in opposite directions like a

---

[m]Remember that a geodesic in a Riemannian manifold (M, g_J) is a path that minimizes the length between any two of its points (provided these are sufficiently close.) The metric area of a surface is a measure of its energy. Thus a minimal surface has minimal energy and is, for example, the shape taken up by a soap film in 3-space that spans a wire frame.

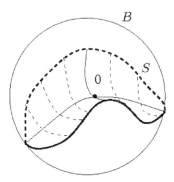

Fig. 3.4.   Any g_0-minimal surface S through 0 has area $\geq \pi$.

saddle and so, unless they are completely flat, have area greater than that of the flat disc.

Families of J-holomorphic curves: Holomorphic (or complex) objects are much more rigid than real ones. For example there are huge numbers of differentiable real valued functions on the 2-sphere S^2, but the only complex (or holomorphic) functions on S^2 are *constant*.[n]

There are infinitely many holomorphic functions if one just asks that they be defined in some small open subset of S^2, but the condition that they be globally defined is very strong. Something very similar happens with complex curves.

If one fixes a point $x \in M$ there are infinitely many real curves through x. In fact there is an infinite dimensional family of such curves, i.e. the set of all such curves can be given the topology of an infinite dimensional space. For real curves it does not matter if we look at little pieces of curves or the whole of a closed curve (e.g. the image of a circle). In the complex case, there still are infinite dimensional families of curves through x if we just look at little pieces of curves. But if we look at closed curves, e.g. maps whose domain is the whole of the 2-sphere, then there is at most a finite dimensional family of such curves. Moreover, Gromov discovered that under many circumstances the most important features of the behavior of these curves does not depend on the precise almost complex structure we are looking at.

---

[n] A well known result in elementary complex analysis is that every bounded holomorphic function that is defined on the whole of the complex plane $\mathbb{C}$ is constant. (These are known as entire functions.) Since the Riemann sphere $S^2 = \mathbb{C} \cup \{\infty\}$ contains $\mathbb{C}$, the same is true for S^2.

For example, if (M, ω) is the complex projective plane with its usual complex structure J_0, then a complex line can be parametrized by a (linear) holomorphic map $f : S^2 \to (M, J_0)$ and so can be thought of as a J_0-holomorphic curve. If we perturb J_0 to some other ω-compatible almost complex structure J, then each complex line perturbs to a J-holomorphic curve. Gromov showed that, just as there is exactly one complex line through each pair of distinct points x, y, there is exactly one of these J-holomorphic curves through each x, y.[o]

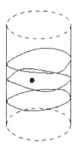

Fig. 3.5. Slicing the cylinder $Z(r)$ with J-holomorphic discs of symplectic area πr^2. In dimension 4, provided that we put on suitable boundary conditions, or better still compactify as explained below, there is precisely one such disc through each point. In higher dimensions, with $Z(r) := D^2(r) \times \mathbb{R}^{2n-2}$, there is at least one.

What we need to prove the nonsqueezing theorem is a related result about cylinders $Z(r)$ as defined in Equation (1).

Slicing cylinders: *Let $\big(Z(r), \omega_0\big)$ be the cylinder in $(\mathbb{R}^4, \omega_0)$, and let J be any ω_0-tame almost complex structure on $Z(r)$ that equals the usual structure outside a compact subset of the interior of $Z(r)$. Then there is a J-holomorphic disc $f : (D^2, \partial D^2) \to (Z(r), \partial Z)$ of symplectic area πr^2 through every point of $Z(r)$.*

Note that here we are interested in discs whose boundary circle ∂D^2 is taken by f to the boundary ∂Z of the cylinder. The above statement is true

---

[o]This very sharp result uses the fact that the complex projective plane has 4 real dimensions. In higher dimensional complex projective spaces, Gromov showed that one can count these curves with appropriate signs and that the resulting sum is one. But now there may be more than one actual curve through two points. Thus the theory is no longer so geometric. The effect is that we know much more about symplectic geometry in 4-dimensions than we do in higher dimensions. However, results like the nonsqueezing theorem are known in all dimensions.

for the usual complex structure J_0. In fact if $w_0 = (z_0, y_0) \in D^2(r) \times \mathbb{R}^2$ then the map

$$f(z) = \left(\frac{z}{r}, y_0 \right) \in D^2(r) \times \mathbb{R}^2$$

goes through the point w_0. Because there is essentially one map of this kind (modulo reparametrizations of f), a deformation argument implies that there always is at least one such map no matter what J we choose.[p]

3.1. *Sketch proof of the nonsqueezing theorem*

Suppose that there is a symplectic embedding

$$\phi : B^4(1) \to Z(r) = D^2(r) \times \mathbb{R}^2.$$

We need to show that $r \geq 1$. Equivalently, by slightly increasing r, we may suppose that the image of the ball lies inside the cylinder, and then we need to show that $r > 1$. We shall do this by using J-holomorphic slices as described above, but where J is chosen very carefully. (Really the whole point of this argument is to choose a J that is related to the geometry of the problem.)

This is how we manage it. In order to make the slicing arguments work we need our J to equal the standard Euclidean structure J_0 near the boundary of $Z(r)$ and also outside a compact subset of $Z(r)$.[q]

But because the image $\phi(B)$ of the ball is strictly inside the cylinder, we can also make J equal to any specified w_0-tame almost complex structure on $\phi(B)$. In particular, we may assume that J equals the pushforward of the standard structure $\phi_*(J_0)$ on $\phi(B)$. In other words, *inside* the embedded ball J is "standard".

Then, the statement above about slicing cylinders says that there is a J-holomorphic disc

$$f : D^2 \to Z(r), \quad f(0) = \phi(0),$$

that goes through the image $\phi(0)$ of the center of the ball and also has boundary on the boundary of the cylinder. Further, the symplectic area of the slice $C = f(D^2)$ is πr^2.

---

[p]To make this argument precise we should partially compactify the target by identifying each boundary circle $\partial D^2(r) \times \{y\}, y \in \mathbb{R}^2$, to a point. The target then becomes $S^2 \times \mathbb{R}^2$. Correspondingly we should look at maps with domain $S^2 = D^2/\partial D^2$. Then the count of J-holomorphic curves through w_0 can be rephrased in terms of the degree of a certain map.

[q]This technical condition is needed so that we can compactify the domain and target as explained earlier.

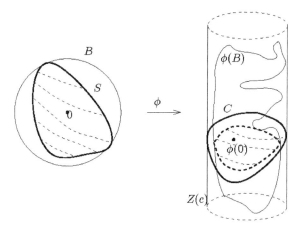

Fig. 3.6.   The g_0-minimal surface S is taken by the embedding ϕ into the intersection (outlined in heavy dots) of the slice C with the image $\phi(B)$ of the ball.

Now consider the intersection $C_B := C \cap \phi(B)$ of the slice with the embedded ball $\phi(B)$. By construction, this goes though the image $\phi(0)$ of the center 0 of the ball B. We now look at this situation from the vantage point of the original ball B. In other words, we look at the inverse image $S := \phi^{-1}(C_B)$ of the curve C_B under ϕ as in Figure 3.6. This consists of all points in the ball that are taken by ϕ into C_B and forms a curve in B that goes through its center. The rest of our argument involves understanding the properties of this curve S in B.

One very important fact is that S is holomorphic in the usual sense of this word, i.e. it is holomorphic with respect to the usual complex structure J_0 on $\mathbb{R}^4 = \mathbb{C}^2$. This follows from our choice of J: by construction, J equals the pushforward of J_0 on the image $\phi(B)$ of the ball, and so, because C_B lies in $\phi(B)$ and is J-holomorphic, it pulls back to a curve S that is holomorphic with respect to the pullback structure J_0.

As we remarked above, this means that S is a minimal surface with respect to the standard metric g_0 on $\mathbb{R}^4$ associated to ω_0 and J_0. So by Equation (3) the area of S with respect to g_0 is at least π. But because S is holomorphic, this metric area is the same as its symplectic area $\omega_0(S)$. This, in turn, equals the ω_0-area of the image curve $\phi(S) = C_B$, because ϕ preserves ω_0.

Finally note that, by construction, C_B is just part of the J-holomorphic slice C through $\phi(0)$. It follows that C_B has strictly smaller ω_0-area than C. (This follows from the taming condition $\omega_0(v, Jv) > 0$ of Equation (1), which, because it is a pointwise inequality, implies that every little piece of a J-holomorphic curve — such as $C \smallsetminus C_B$ — has strictly positive symplectic area.) But our basic theorem about slices says that $\omega_0(C) = \pi r^2$. Putting this all together, we have the following string of inequalities and equalities:

$$\pi \leq g_0\text{-area } S = \omega_0\text{-area } S = \omega_0\text{-area } \phi(S) < \omega_0\text{-area } C = \pi r^2.$$

Thus $\pi < \pi r^2$. This means that $r > 1$, which is precisely what we wanted to prove.

Acknowledgements

I wish to thank Yael Degany and James McIvor for their useful comments on an earlier draft of this note.

References

1. V. I. Arnold, *Mathematical methods in Classical Mechanics* Springer (1978).
2. A. Cannas, *Lectures on Symplectic Geometry*, Lecture Notes in Mathematics vol 1764, 2nd ed. Springer (2008).
3. K. Cieliebak, H. Hofer, J. Latschev and F. Schlenk, Quantitative symplectic geometry, arXiv:math/0506191, *Dynamics, Ergodic Theory, Geometry MSRI*, **54** (2007), 1–44.
4. M. Gromov, Pseudo holomorphic curves in symplectic manifolds, *Inventiones Mathematicae*, **82** (1985), 307–47.
5. C. Marle, The inception of Symplectic Geometry: the works of Lagrange and Poisson during the years 1808–1810, arxiv:math/0902.0685.
   An article about the early history of the idea of a symplectic form. A more standard description of the history can be found in [1].
6. D. McDuff, Symplectic structures – a new approach to geometry, *Notices of the Amer. Math. Soc.* **45** (1998), 952–960.
   This is slightly more technical than the current article, but is still quite accessible.
7. D. McDuff, A glimpse into symplectic geometry, in *Mathematics: Frontiers and Perspectives 2000*, AMS, Providence.
   Another survey article that emphasizes the two-fold nature of symplectic geometry.
8. D. McDuff, Floer Theory and Low dimensional Topology, *Bull. Amer. Math. Soc.* (2006).
   A survey article describing some recent applications of symplectic geometry to low dimensional topology.
9. D. McDuff and D. Salamon, *Introduction to Symplectic Topology*, 2nd edition (1998) OUP, Oxford, UK.
10. D. McDuff and D.A. Salamon, *J-holomorphic curves and symplectic topology*. Colloquium Publications **52**, American Mathematical Society, Providence, RI, (2004).

The two above books are comprehensive (and not very elementary) introductions to the field.

11. D. McDuff, Symplectic embeddings of 4-dimensional ellipsoids, to appear in *Journ. of Top.* (2009).

12. D. McDuff and F. Schlenk, The embedding capacity of 4-dimensional symplectic ellipsoids, in preparation.

13. J. Milnor, *Topology from a differentiable viewpoint.*
An elementary and very clear introduction to this fascinating subject.

14. E. Opshtein, Maximal symplectic packings of $\mathbb{P}^2$, arxiv:0610677.

REGULAR PERMUTATION GROUPS AND CAYLEY GRAPHS

CHERYL E. PRAEGER

School of Mathematics and Statistics
University of Western Australia
Crawley WA 6009
Australia

Regular permutation groups are the 'smallest' transitive groups of permutations, and have been studied for more than a century. They occur, in particular, as subgroups of automorphisms of Cayley graphs, and their applications range from obvious graph theoretic ones through to studying word growth in groups and modeling random selection for group computation. Recent work, using the finite simple group classification, has focused on the problem of classifying the finite primitive permutation groups that contain regular permutation groups as subgroups, and classifying various classes of vertex-primitive Cayley graphs. Both old and very recent work on regular permutation groups are discussed.

Keywords: Permutation groups; Cayley graphs.

1. Introduction

Finite primitive permutation groups containing a regular subgroup have been studied for more than one hundred years, while the theory of permutation groups is even older, going back to the origins of Group Theory in the early nineteenth century. Cayley graphs encode the structure of a group, and are a central tool in combinatorial and geometric group theory. In this chapter I will introduce these concepts, discuss some of their history and applications, and summarise some very recent classification results concerning finite regular permutation groups and Cayley graphs the proofs of which rely on the finite simple group classification.

1.1. *Permutation groups and regularity*

A *permutation* of a set Ω is a bijection $g : \Omega \to \Omega$, and the set of all permutations of Ω forms the *symmetric group* $\mathrm{Sym}(\Omega)$ under composition.

For example, performing the permutation $g = (1, 2)$ of $\Omega = \{1, 2, 3\}$, followed by $h = (2, 3)$, yields the permutation $gh = (1, 3, 2)$. By a permutation group on Ω we mean an arbitrary subgroup of $\text{Sym}(\Omega)$. Permutation groups provide a basic measure of symmetry of a system and so have many important applications, for example in Graph Theory (as automorphism groups), Geometry (as groups of collineations), in Number Theory and Cryptography (as Galois groups and groups associated with elliptic curves), and in Differential Equations (where the nature of the solutions depends on the symmetries of the equation).

A permutation group $G \leq \text{Sym}(\Omega)$ is said to be *transitive* if all points of Ω are equivalent under elements of G, that is, for any $\alpha, \beta \in \Omega$, some element of G maps α to β. If G is finite then the number of permutations in G that fix any given point of Ω is equal to $|G|/|\Omega|$ (where $|G|$, $|\Omega|$ denote the cardinality of G and Ω respectively). A transitive group G on Ω is *regular* if only the identity element fixes a point of Ω. In the finite case this is equivalent to $|G| = |\Omega|$, and thus regular permutation groups are the 'smallest possible transitive' groups. Here is a small concrete example.

Example 1.1. Let $G = \langle (0, 1, 2, 3, 4) \rangle$ on $\Omega = \{0, 1, 2, 3, 4\}$; or alternatively, regarding Ω as the set of integers modulo 5, then G is the group $\mathbb{Z}_5$ acting by addition modulo 5. The group G is transitive on Ω and $|G| = |\Omega| = 5$. Thus G is regular.

This is an example of the following general construction.

Example 1.2. Let G be a group, and set $\Omega := G$. For each $g \in G$ define $\rho_g : \Omega \rightarrow \Omega$ by $x \mapsto xg$ for $x \in \Omega$, and note that ρ_g is a bijection so $\rho_g \in \text{Sym}(\Omega)$. The set $G_R = \{\rho_g | g \in G\}$ is a regular permutation group on Ω and is isomorphic to G. It is called the *right regular representation* of G (since elements act by multiplication on the right).

1.2. Cayley graphs

A natural way to view diagramatically the right regular representation of a group G described in Example 1.2 is by representating the elements of $\Omega = G$ as vertices of a graph, so that the action of ρ_g for an element $g \in G$ is represented by a collection of directed edges, one for each $b \in G$ drawn from b to bg, see Figure 1.2.

The collection of directed edges representing $\rho_{g^{-1}}$ is the same as that for ρ_g but with the arrows going in the opposite direction on each edge. We can represent on the same diagram the edges corresponding to the maps ρ_s

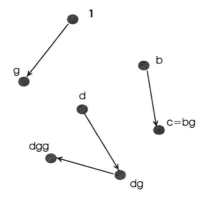

Fig. 1.1.   Action of ρ_g for Example 1.2.

for all s in a subset S of G, and if S is closed under inverses we can remove the arrows indicating direction and obtain an undirected 'Cayley graph' of G relative to S. This graph is connected provided the subset S generates the group G, a condition we will assume in the definition.

Definition 1.1. For a generating set S of a group G such that $s \in S$ if and only if $s^{-1} \in S$, the *Cayley graph* for G relative to S is the graph with vertex set $\Omega = G$, and edges $\{g, sg\}$ for $g \in G, s \in S$. It is denoted $\mathrm{Cay}(G, S)$.

Each ρ_g leaves the edge set of $\mathrm{Cay}(G, S)$ invariant (as a set), and so is an automorphism of this Cayley graph. Thus G_R is a subgroup of the automorphism group $\mathrm{Aut}(\mathrm{Cay}(G, S))$, and is regular on the vertex set. In particular $\mathrm{Cay}(G, S)$ is vertex-transitive.

Example 1.3. For $G = \mathbb{Z}_5$, and $S = \{1, 4\}$, we obtain $\Gamma = \mathrm{Cay}(G, S) \cong C_5$, with automorphism group $\mathrm{Aut}(\Gamma) = D_{10}$, see Figure 1.2.

Cayley graphs are named after the nineteenth century English mathematician Arthur Cayley (1821-1895). They are important in combinatorics, statistical designs, and computation. For example, the special type of Cayley graphs called circulant graphs are used in experimental layouts for statistical experiments and for many constructions in combinatorics. An important early explicit construction of an infininte family of 'expander graphs' by Lubotzky, Phillips, and Sarnak [20] in 1988 (the Ramanujan graphs) produced Cayley graphs for a family of simple groups. Cayley

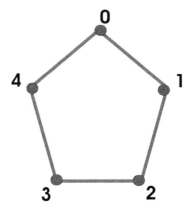

Fig. 1.2.   Cayley graphs for Example 1.3.

graphs are also used as models to analyse the construction of approximately random group elements in randomised group algorithms, see [1].

In the remainder of the chapter we discuss various aspects of regular permutation groups and Cayley graphs:

* Recognising Cayley graphs
* Primitive Cayley graphs
* B-groups
* Exact group factorisations

2. A recognition problem for Cayley graphs

Some Cayley graphs possess many additional automorphisms, and admit constructions that give no hint that they are in fact Cayley graphs. A famous example of this is the *Higman–Sims graph* $\Gamma = \Gamma(HS)$ which has 100 vertices and valency 22. A construction of this graph by D. G. Higman and C. C. Sims in 1967 was instrumental in their discovery of the sporadic simple group now called the *Higman–Sims* simple group HS. The automorphism group of the graph is $A := \mathrm{Aut}(\Gamma) = \mathrm{HS}.2$, and its construction by Higman and Sims used the important fact that a vertex stabiliser is the sporadic almost simple Mathieu group $A_\alpha = M_{22}.2$, and has orbits related to the associated Steiner system $S(3, 6, 22)$. From its construction it is not obvious that $\Gamma(HS) = \mathrm{Cay}(G, S)$ for the $G = (Z_5 \times Z_5) : [4]$. The graph HS was constructed before the work of Higman and Sims by Dale Mesner [24] in his PhD thesis, but Mesner did not examine the automorphism group of this graph.

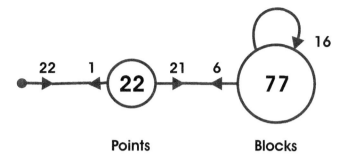

Fig. 2.1. Distance diagram for the Higman–Sims graph.

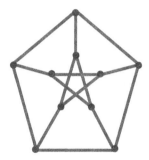

Fig. 2.2. The Petersen graph.

Although all Cayley graphs are vertex-transitive, not all vertex-transitive graphs are Cayley graphs. The smallest such example is the Petersen graph P shown in Figure 2.2. The crucial property, that determines whether a given vertex-transitive graph is a Cayley graph, is the presence of a regular subgroup of automorphisms.

Theorem 2.1. *A graph* Γ *is a Cayley graph if and only if there exists a subgroup* $R \le \mathrm{Aut}(\Gamma)$ *with* R *regular on vertices. In this case* $\Gamma \cong \mathrm{Cay}(R, S)$ *for some* S.

In the case of the Petersen graph $\Gamma = P$, the automorphism group is $\mathrm{Aut}(\Gamma) = S_5$, and this has no regular subgroup of order 10, since every involution (element of order 2) in $\mathrm{Aut}(\Gamma)$ fixes a vertex, while an involution in a regular subgroup would have no fixed points. Theorem 2.1 suggests that, in order to decide if Γ is a Cayley graph, one should first determine $\mathrm{Aut}(\Gamma)$, and then search for a regular subgroup R. Both of these steps are difficult to carry out in general.

On the other hand, exhaustive enumeration of vertex-transitive graphs of small orders suggests that Cayley graphs are 'common' among vertex-transitive graphs. For example, there are 15,506 vertex-transitive graphs with 24 vertices, and of these, 15,394 are Cayley graphs [26]. Based on the evidence of such enumerations it has been conjectured, see [21, 22], that most finite vertex-transitive graphs are indeed Cayley graphs.

Conjecture 2.1 (McKay-Praeger). *As $n \to \infty$*

$$\frac{Number \ of \ Cayley \ graphs \ on \leq n \ vertices}{Number \ of \ vertex\text{-}transitive \ graphs \ on \leq n \ vertices} \to 1$$

Various suggestions have been made regarding a way forward to understand the 'vertex-transitive/Cayley graph question'. One of these is the *Non-Cayley Project*: determine all positive integers n such that all vertex-transitive graphs on n vertices are Cayley graphs. Examples include all prime powers p^i for $i = 1, 2, 3$. This problem was posed by Dragan Marušic [23] in 1983, and much progress has been made, see for example [21, 22]. A second suggestion, by Ming Yao Xu [29] in 1998 was to study *normal Cayley graphs*, that is, Cayley graphs $\Gamma = \text{Cay}(G, S)$ for which $G_R \lhd \text{Aut}(\Gamma)$; for these graphs it is easy to recognise that they are Cayley graphs since the regular subgroup G_R is normal in the full automorphism group.

A third strand of research, and *the one which I will discuss in this chapter*, is to study the *primitive Cayley graphs*. These are the Cayley graphs $\Gamma = \text{Cay}(G, S)$ for which the full automorphism group is vertex-primitive, that is to say, the only vertex-partitions invariant under $\text{Aut}(\Gamma)$ are the trivial ones with either just one part, or all parts of size one. To give a context for this condition, note that each subgroup $H < G$ corresponds to a G_R-invariant vertex-partition of Γ, namely the set of right H-cosets in G, and each G_R-invariant vertex-partition has this form. Thus in order for Γ to be vertex-primitive, we need, for each proper non-trivial subgroup H of G, some additional automorphism *not preserving* the H-coset partition. In other words, if G has many subgroups, we need correspondingly many additional partition-breaking automorphisms for a primitive Cayley graph.

Nevertheless, for each group G there is at least one Cayley graph for G that is vertex-primitive: take $S = G \setminus \{1_G\}$, and then $\text{Cay}(G, S)$ is the *complete graph K_n* on $n = |G|$ vertices with automorphism group the symmetric group S_n which is primitive on the n vertices. We are interested to know about *non-complete primitive Cayley graphs*. One example is the Higman-Sims graph discussed above.

3. Cayley graphs and *B*-groups

In a purely group-theoretic setting, in 1911, the English mathematician W. Burnside discovered that the presence of a cyclic regular subgroup of non-prime, prime power order, in a primitive permutation group forces the group to be doubly transitive, a very strong condition (namely, all ordered point-pairs are equivalent under the group action). Burnside believed, but did not prove, that the same was true for all regular abelian groups other than the elementary abelian ones. The eminent German mathematician Issai Schur published a paper on this problem in 1933, generalising Burnside's results significantly. Developments of Schur's methods led to the theory of Schur rings that in turn led to Hecke algebras, which are important in representation theory.

In the language of Cayley graphs, Burnside's result is equivalent to the statement: *the only primitive Cayley graphs for cyclic groups of non-prime, prime power, order are complete graphs.* This is because a 2-transitive action by a graph automorphism group on the vertices of a connected graph is possible only if the graph is a complete graph. Schur's work yielded a generalisation of this assertion, namely his work showed that the adjective 'prime power' could be removed. Further generalisations of Burnside's result were obtained in 1935 and 1950 by Helmut Wielandt for abelian groups, and dihedral groups, respectively (see [28, Section 25] for a short account).

In 1955, Wielandt coined the term *B-group* for a group G of order n if (using the language of Cayley graphs) the only primitive Cayley graph for G is K_n. (See [28, Section 25].) Thus many abelian groups, and in particular most cyclic groups, as well as all dihedral groups, were at that time known to be *B*-groups. Moreover, 2-transitive permutation groups were of great interest in the succeeding decades, as several families of simple groups were discovered as 2-transitive permutation groups (the Suzuki groups and the Ree groups). Indeed various studies during the 1960's and 1970's focused on identifying further families of *B*-groups.

Recent work on primitive Cayley graphs produces results relying on very deep mathematics, in the sense that the proofs rely on the classification of the finite simple groups. Modern work in the permutation group setting focuses on the following problem.

Problem 3.1. *Find all pairs* (G, H), *such that* $G < H \le \mathrm{Sym}(\Omega)$, *with* G *regular and* H *primitive (and not 2-transitive) on the finite set* Ω.

In this context, we may identify Ω with the set G in such a way that G acts by right multiplication, and then, taking S to be any self-inverse proper subset of $G \setminus \{1\}$, we have H a vertex-primitive subgroup of $\mathrm{Aut}(\Gamma)$ where $\Gamma = \mathrm{Cay}(G, S)$ is a primitive Cayley graph for G, not a complete graph. *The aims of these investigations are to understand which primitive groups H arise, and to understand better the associated primitive Cayley graphs Γ.*

One reason for this shift of focus is that, as a consequence of the finite simple group classification, all the finite 2-transitive permutation groups are now known, (see for example [6, Chapter 7.7] for a description of the examples).

4. A fascinating density result

The following beautiful theorem of Cameron, Neumann, and Teague [5] provides some important information on the possible orders of B-groups. The theorem dates from 1982, soon after the announcement of the finite simple group classification, and Theorem 4.1 gives an informal version of it. In Theorem 4.1, a certain property associated with a positive integer n is asserted to hold for 'almost all n'. This phrase is used in a number-theoretic sense, namely, as a real number x approaches infinity, the proportion of integers $n \leq x$ for which the property holds approaches the limit 1.

Theorem 4.1. *For almost all positive integers n, the only primitive permutation groups on a set of size n are A_n and S_n.*

Here S_n denotes the finite symmetric group $\mathrm{Sym}(\Omega)$ consisting of all permutations of the set $\Omega := \{1, \ldots, n\}$, and A_n is the alternating group comprising all even permutations of Ω; A_n has index 2 in S_n. To give a more precise version of Theorem 4.1, for a positive real number x, let $N(x)$ denote the number of positive integers $n \leq x$ such that there exists a primitive subgroup H of S_n with $H \neq A_n, S_n$. It is proved in [5] that

$$N(x) = 2\pi(x) + O(x^{1/2})$$

where $\pi(x)$ is the number of primes less than x. Thus $N(x)/x \sim 2x/\log(x)$ and in particular $N(x)/x \to 0$ as $x \to \infty$.

This implies in particular that for 'almost all' positive integers n and for each group G of order n, if $G < H \leq S_n$ with G regular and H primitive, then $H = A_n$ or S_n, and hence H is 2-transitive. For every integer n with this property, every group of order n is a B-group, as defined in Section 3.

We are interested in primitive Cayley graphs that are not complete graphs, or equivalently, in primitive permutation groups on n points that are not 2-transitive and that contain a regular subgroup. By Theorem 4.1, the subset of all such integers n has density zero in the natural numbers. Thus perhaps there may be something special about the groups that are not B-groups.

5. Exact factorisations of groups

Each Cayley graph for G admtting a vertex-primitive subgroup H of automorphisms gives rise to a special kind of group factorisation. Wielandt's condition involves a group inclusion $G < H \le \mathrm{Aut}(\Gamma)$ with G regular and H primitive on vertices, where $\Gamma = \mathrm{Cay}(G, S)$ for some S. Setting K to be the stabiliser in H of a vertex of Γ, we have

$$H = GK \text{ and } G \cap K = 1 \text{ with } K \text{ a maximal core-free subgroup of } H. \quad (1)$$

An expression $H = GK$ with $G \cap K = 1$ is called an *exact factorisation* of H. The subgroup K is *core-free*, that is its core $\cap_{h \in H} K^h$ is the identity subgroup, since $\cap_{h \in H} K^h$ fixes all the vertices of Γ. Also the maximality of K in H holds because H is vertex-primitive. On the other hand, for each exact factorisation $H = GK$ with K a maximal core-free subgroup, we may take Γ to be any graph with vertices the right cosets of K in H, and with edges the pairs $\{Kx, Ky\}$ for which xy^{-1} lies in a specified inverse-closed union of K-double cosets (that is, subsets of the form KgK). Such a graph admits H, acting by right multiplication, as a vertex-primitive group of automorphisms. Moreover, since $G \cap K = 1$, G acts regularly on vertices, and it follows from Theorem 2.1 that $\Gamma \cong \mathrm{Cay}(G, S)$ for some S. Every finite group G occurs at least twice with appropriate groups H, K in (1), namely with $H = A_n$ or S_n and $K = A_{n-1}$ or S_{n-1} respectively, where $|G| = n$, taking G to act regularly. However in these cases the only possible Cayley graph is the complete graph K_n (see the last paragraph of Section 2). Problem 3.1 at the end of Section 3 is equivalent to the following.

Problem 5.1. *For a finite group H, find all exact factorisations $H = GK$ with K maximal and core-free in H, and such that $(H, K) \neq (A_n, A_{n-1})$ or (S_n, S_{n-1}), where $n = |G|$.*

This group factorisation problem is not new, but modern methods have led to almost complete solutions for some classes of primitive groups H, and some classes of regular groups G. As an example, consider exact factorisations of the finite alternating and symmetric groups. In 1935, G. A. Miller [25] studied exact factorisations $H = GK$ of $H = A_n$. He gave interesting examples of exact factorisations for some values of n, and also he gave examples of integers n for which the only exact factorisations have $K = A_{n-1}$. More than forty years later in 1980, Wiegold and Williamson [27] were able to classify all factorisations $H = GK$ with $H = A_n$ or S_n, and we can deduce from their work an explicit list of all triples (H, K, G) satisfying the conditions of Problem 5.1 with $H = A_n$ or S_n for some n, (see [3] or [19] for details).

Similarly complete classifications of exact factorisations were obtained for the other non-classical almost simple groups H by examining classifications of all factorisations $H = GK$ of such groups H. A finite group H is *almost simple* if H has a unique minimal normal subgroup, denoted $\mathrm{Soc}(H)$, and $\mathrm{Soc}(H)$ is a finite nonabelian simple group. The factorisations of the finite exceptional almost simple groups of Lie type were classified in 1987 by Hering, Liebeck and Saxl [11], and it is clear from the short list of these factorisations that none of them is exact. More recently the factorisations of the 26 sporadic almost simple groups were classified by Giudici [9], building on the classification of the maximal factorisations of these groups in [16]. Working from this list one can obtain the list of exact factorisations for the sporadic groups. Thus, as with many problems about finite simple groups, solving Problem 5.1 in the case of almost simple groups is reduced quickly to the case of finite classical groups (provided we invoke the finite simple group classification that says that the classical groups are the only remaining cases). We discuss the resolution of this case in Section 7.

We have expressed the 'primitive Cayley graph problem' in the language of permutation groups (as Problem 3.1) and abstract groups (as Problem 5.1). This repeated articulation of the problem has not been merely a translation exercise. It has led to applications beyond the realms of groups and graphs. For example, exact factorisations have been used for constructing semisimple Hopf algebras, see [7], though the construction given there using bicrossproducts goes back to Kac and Takeuchi.

6. Primitive Cayley graphs for various groups G

As soon as the finite simple group classification seemed imminent. many accounts appeared exploring its far-reaching consequences. As part of a very interesting paper [8] of this type, in 1980, Walter Feit recorded the list of all finite 2-transitive permutation groups on n points, other than the alternating and symmetric groups A_n and S_n, that contain a regular cyclic subgroup. The degrees n that occur are $11, 23$ and numbers of the form $(q^k - 1)/(q - 1)$ for prime powers q. This gave further insight into the situation considered by Burnside and discussed in Section 3.

Some twenty years later in [12] Gareth Jones classified all the finite primitive permutation groups H containing a regular cyclic subgroup. Extending this classification in [13, 14], Cai Heng Li classified all the finite primitive permutation groups H containing a regular abelian or dihedral subgroup. These three classifications had important consequences for studying various classes of graphs and in particular of embeddings of Cayley graphs in orientable and non-orientable surfaces. For example, the work in [13, 14] led to a determination of all 2-arc-transitive Cayley graphs of abelian groups, all symmetric circulant graphs, and all rotary Cayley maps of simple groups.

Other families of groups G have been studied recently. For example, significant inroads into solving Problem 5.1 have been made for groups G of square-free order by Li and Seress in [15]. Perhaps the most interesting family of insoluble groups G for which a complete solution to Problem 5.1 has been achieved is the family of finite almost simple groups. The solution for the the the alternating, sporadic, and exceptional Lie type cases was discussed in Section 5, leaving the classical groups as the major class outstanding. Some infinite families of classical groups were dealt with by Baumeister in [2, 3], and a complete classification has been achieved in work yet to be published by Liebeck, Saxl and the author in [19]. The most important resource used in the analysis is the classification of the maximal factorisations of the finite almost simple groups in [16, 17].

We describe briefly the results in [19] for almost simple groups G. Every finite nonabelian simple group G occurs as a regular subgroup of a finite primitive group H of diagonal type, and hence the finite simple groups are not B-groups. For the primitive Cayley graphs $\mathrm{Cay}(G, S)$ corresponding to such 'diagonal embeddings' of G, the generating set S is a union of conjugacy classes of G. It turns out, see [19, Theorem 1.6], that the only simple groups G that embed into primitive groups H which are not of diagonal type and are not 2-transitive, are certain of the alternating groups. Thus the primitive Cayley graphs for finite nonabelian simple groups are essentially well understood.

Theorem 6.1. *Let* $\mathrm{Cay}(G, S)$ *be a primitive Cayley graph for a finite nonabelian simple group* G. *Then either* S *is a union of conjugacy classes of* G, *or* $G = A_{p^2-2}$ *for some prime* $p \equiv 3 \pmod 4$.

In contrast to this (see [19, Theorem 1.4 and Corollary 1.5]), if G is almost simple but not simple, that is, if $G \neq \mathrm{Soc}(G)$, then there are very few possibilities for both the almost simple regular group G and the primitive permutation group H in Problem 5.1. In particular, the non-simple almost simple groups are all B-groups, except for S_{p-2} (p prime), $L_2(16).4$, and $L_3(4).2$.

Theorem 6.2. *Suppose that G, H, K are as in Problem 5.1 with G almost simple and $G \neq \mathrm{Soc}(G)$. Then also H is almost simple, and $G, \mathrm{Soc}(H)$, and $\mathrm{Soc}(H) \cap K$ are as in one of the rows of Table 6.1.*

Table 6.1.   Almost simple regular subgroups.

G	$\mathrm{Soc}(H)$	$\mathrm{Soc}(H) \cap K$
S_{p-2} ($p \geq 7$ prime)	A_p, A_{p+1}	$p.\binom{p-1}{2}$, $L_2(p)$ (resp.)
S_5	A_9 $\mathrm{Sp}_4(4)$ $\mathrm{Sp}_6(2)$ $O_8^+(2)$ $\mathrm{Sp}_8(2)$	$L_2(8).3$ $L_2(16).2$ $G_2(2)$ $O_7(2)$ $O_8^-(2)$
$L_2(16).4$	$\mathrm{Sp}_6(4)$, $O_8^+(4)$	$G_2(4)$, $O_7(4)$ (resp.)
$L_3(4).2$	M_{23}, M_{24}	23.11, $L_2(23)$ (resp.)

7. Types of finite primitive groups

When studying almost any problem concerning finite primitive permutation groups these days, one usually considers the problem separately for the various 'types' of primitive groups described by the O'Nan Scott Theorem, see for example [6, Chapter 4]. There are several versions of the subdivision of primitive groups provided by this theorem, and each has its uses in appropriate applications. For several types of finite primitive groups, a regular subgroup is evident from the type definition, but even for groups of these types there may exist other regular subgroups that are not at all obvious. For other primitive types, it is not clear at all when groups of these types contain regular subgroups.

For example, each finite primitive group of affine type is a group of affine transformations of a finite vector space, and the subgroup of translations acts regularly. However there are sometimes additional regular subgroups in these groups, and examples are given in [10]. Similarly, groups of diagonal type always contain regular subgroups that are products of simple subnormal subgroups. They also may contain additional regular subgroups provided their simple subnormal subgroups admit a nontrivial factorisation. A thorough study of this issue was conducted in 2000 by Liebeck, Saxl and the author [18]. It gave a satisfactory description of the regular subgroups of primitive permutation groups for all but two of the types of primitive groups. For one of these types, the product action type, the problem remains unsolved.

The other primitve type consists of the almost simple primitive groups H. In Section 5 we described briefly how the regular subgroups were classified in all the almost simple primitive groups H except the case where H is a finite classical group.

8. Exact factorisations of finite classical groups

A finite almost simple classical group H either contains a projective group $\mathrm{PSL}(n, q)$, or is an n-dimensional symplectic, unitary or orthogonal group defined over a finite field $\mathbb{F}_q$ of order q. Exact factorisations for the primitive representations of unitary groups and the 8-dimensional orthogonal groups were classified by Baumeister in [2, 3]. A complete solution for all finite classical groups is given by Liebeck, Saxl and the author in the monograph [19]. Finding all the exact factorisations for the finite classical groups is indeed the heart of Problems 3.1 and 5.1 for almost simple groups H.

The full classification of regular subgroups of finite primitive classical groups is given in [19, Theorem 1.1], and is summarised in a whole-of-page table in [19, Section 16]. The detailed analysis required to obtain this result produced subsidiary results of independent interest. To explain why this might be a natural outcome, consider the (primitive) action of a finite classical group H on, say, totally isotropic or non-singular subspaces, of a given dimension, of the underlying vector space, and let K be the stabiliser in H of such a subspace. For a subgroup G of H, a factorisation $H = GK$ occurs if and only if G is transitive on the set of such subspaces. The approach in [19, Section 4] is to produce several detailed results determining the subgroups of classical groups which are transitive on these types of subspaces. These results should have useful applications in future geometrical and combinatorial investigations. For the current problem, these lists of transitive subgroups, for subspace actions of classical groups, were examined in detail to decide which of them corresponded to an exact factorisation.

Another kind of primitive action of finite classical groups is on so-called antiflags of the underlying geometry, and as preparation for handling the cases where the action of H is on antiflags, the work in [19, Section 3] gives a classification of antiflag transitive linear groups. This updates and slightly generalizes the famous theorem of Cameron and Kantor [4] about such groups.

Despite the fairly long lists of examples of almost simple primitive permutation groups with a regular subgroup, there are essentially only four infinite families of groups that occur as regular subgroups. As recorded in [19, Corollary 1.2], if $n > 3 \cdot 29!$ to avoid the finitely many exceptions, and if G is a regular subgroup of a primitive group H on $\{1, \ldots, n\}$, such that $H \neq A_n, S_n$, then one of the following holds.

(i) G is metacyclic, of order $(q^r - 1)/(q - 1)$ for some prime power q;
(ii) G is a subgroup of odd order $q(q-1)/2$ of a 1-dimensional affine group $A\Gamma L_1(q)$ for some prime power $q \equiv 3 \pmod 4$;
(iii) $G = A_{p-2}$ or S_{p-2} (p prime), or $A_{p-2} \times 2$ (p prime, $p \equiv 1 \pmod 4$));
(iv) $G = A_{p^2-2}$ (p prime, $p \equiv 3 \pmod 4$).

It would be very interesting to solve the remaining problem of determining the regular subgroups of finite primitive groups in product action, as this would give us a complete picture of the *infinite families* of groups G that occur as regular subgroups of finite primitive permutation groups.

References

1. L. Babai, Local expansion of vertex-transitive graphs and random generation in finite groups, *Proceedings of the twenty-third annual ACM symposium on Theory of computing*, New Orleans, Louisiana, 1991, pp. 164–174.
2. B. Baumeister, Primitive permutation groups of unitary type with a regular subgroup, *Bull. Belg. Math. Soc. Simon Stevin* **12** (2005), 657–673.
3. B. Baumeister, Primitive permutation groups with a regular subgroup, *J. Algebra* **310** (2007), 569–618.
4. P.J. Cameron and W.M. Kantor, 2-transitive and antiflag transitive collineation groups of finite projective spaces, *J. Algebra* **60** (1979), 384–422.
5. P.J. Cameron, P.M. Neumann and D.N. Teague, On the degrees of primitive permutation groups, *Math. Z.* **180** (1982), 141–149.
6. J.D. Dixon and B. Mortimer, *Permutation groups*, Springer-Verlag, Graduate Texts in Mathematics **163**, New York, 1996.
7. P. Etingof, S. Gelaki, R. M. Guralnick and J. Saxl, Bi-perfect Hopf algebras, *J. Algebra* **232** (2000), 331–335.
8. W. Feit. Some consequences of the classification of finite simple groups. In *The Santa Cruz conference on finite groups*, Proc. Sympos. Pure Math. **37** (American Mathematical Society, 1980), pp. 175–181.
9. M. Giudici, Factorizations of sporadic simple groups, *J. Algebra* **304** (2006), 311–323.

10. P. Hegedüs, Regular subgroups of the affine group, *J. Algebra* **225** (2000), 740–742.
11. C.Hering, M.W. Liebeck and J. Saxl, The factorizations of the finite exceptional groups of Lie type, *J. Algebra* **106** (1987), 517–527.
12. G.A. Jones, Cyclic regular subgroups of primitive permutation groups, *J. Group Theory*, **5** (2002), 403–407.
13. C.H. Li, The finite primitive permutation groups containing an abelian regular subgroup, *Proc. London Math. Soc.* **87** (2003), 725–747.
14. C.H. Li, Finite edge-transitive Cayley graphs and rotary Cayley maps, *Trans. Amer. Math. Soc.* **358** (2006), 4605–4635.
15. C.H. Li and A. Seress, On vertex-transitive non-Cayley graphs of square-free order, *Designs, Codes and Cryptography* **34** (2005), 265–281.
16. M.W. Liebeck, C.E. Praeger and J. Saxl, The maximal factorisations of the finite simple groups and their automorphism groups, *Mem. Amer. Math. Soc.* **86** (1990), no. 432.
17. M.W. Liebeck, C.E. Praeger and J. Saxl, On factorizations of almost simple groups, *J. Algebra* **185** (1996), 409–419.
18. M.W. Liebeck, C.E. Praeger and J. Saxl, Transitive subgroups of primitive permutation groups, *J. Algebra* **234** (2000), 291–361.
19. M.W. Liebeck, C.E. Praeger and J. Saxl, On regular subgroups of primitive permutation groups, *Memoirs Amer. Math. Soc.* to appear.
20. A. Lubotzky, R. Phillips, and P. Sarnak, Ramanujan graphs. *Combinatorica* **8** (1988), 261–277.
21. B. D. McKay and C. E. Praeger, Vertex-transitive graphs which are not Cayley graphs, I, *J. Austral. Math. Soc.(A)* **56** (1994), 53–63.
22. B. D. McKay and C. E. Praeger, Vertex-transitive graphs that are not Cayley graphs, II, *J. Graph Theory* **22** (1996), 321–334.
23. D. Marušic, Cayley properties of vertex symmetric graphs. *Ars Combin.* **16** (1983), B, 297–302.
24. Dale Marsh Mesner, *An investigation of certain combinatorial properites of partially balanced incomplete block experimental designs and association schemes, with a detailed study of designs of Latin square and related types,* PhD Thesis, Michigan State University, 1956.
25. G. A. Miller, Groups Which are the Products of Two Permutable Proper Subgroups, *Proceedings of the National Academy of Sciences* **21** (1935), 469–472.
26. C. E. Praeger and G. Royle, Constructing the vertex transitive graphs of order 24, *J. Symbolic Computation* **8** (1989), 309–326.
27. J. Wiegold and A. G. Williamson, The factorisation of the alternating and symmetric groups, *Math. Z.* **175** (1980), 171–179.
28. H. Wielandt, *Finite Permutation Groups*, Academic Press, 1964.
29. Ming Yao Xu, Automorphism groups and isomorphisms of Cayley digraphs, *Discrete Math.* **182** (1998), 309–319.

ARITHMETIC OF ELLIPTIC CURVES
THROUGH THE AGES

R. SUJATHA

School of Mathematics
Tata Institute of Fundamental Research
Homi Bhabha Road, Colaba
Mumbai, INDIA 400005
sujatha@math.tifr.res.in

This expository article is based on a talk that was given at the EWM Symposium held at Cambridge, U.K., in October 2007. The talk was aimed at a broad and general audience and I have tried to retain the flavour of the original lecture while converting it to its present text version. I have also attempted to make the bibliography as comprehensive as possible, but given the vastness of the subject, apologise for any inadvertent omissions. I would like to thank the organisers of the EWM conference for the invitation to speak, and John Coates for helpful discussions and comments. It is a pleasure to thank Chennai Mathematical Institute for hospitality accorded both at the time of preparing the talk, and later, while writing the article.

1. Introduction

The human mind has long contemplated the problem of solving cubic equations. A Babylonian clay tablet from around 1700 B.C., presently exhibited at the Berlin museum is perhaps the oldest piece of evidence in this direction. It lists many problems, some of which can be translated in modern mathematical language, to solving degree three polynomial equations in one variable [19]. Many centuries later, the Greeks, especially Diophantus, were concerned with rational and integral solutions of these equations. While the problem of solving cubic equations in one variable was settled by the 16th century, due to the efforts of del Ferro, Cardano, Tartaglia, Viète and others, mathematicians like Fermat, Euler, Lagrange began to uncover the deep arithmetical mysteries of cubic curves in the 17th and 18th century. In another direction, elliptic integrals arose from the study of the arc lengths of an ellipse, and the theory of elliptic equations grew out of this. We owe to Fermat the discovery of the procedure of infinite descent (see [32]), which he used to prove that $x^4 + y^4 = 1$ has no solution in the field $\mathbb{Q}$ of rational numbers with $xy \neq 0$. He also pondered, leaving no written traces of any success, about the rational solutions of the equation $x^3 + y^3 = 1$. Non-singular cubic curves are the first non-trivial examples of projective curves. For an excellent historical survey of these subjects, see Weil [37]. In more recent times, the study of elliptic curves (see §2) has connections with areas as diverse as complex topology, algebraic geometry and of course, number theory. Some of the most striking unsolved problems of number theory are concerned with the study of rational points (that is, points with coordinates in $\mathbb{Q}$) on elliptic curves. One of our broad aims in this article is to give an idea of how they provide a common ground for ancient and modern themes in number theory.

2. Elliptic curves and number theory

Algebraic curves are the simplest objects of study in algebraic geometry. Projective algebraic curves are classified, upto birational transformations, by a basic birational invariant, called the genus (see [17]). If the curve is a non-singular plane curve of degree d, then the genus is given by $(d - 1)(d - 2)/2$. An elliptic curve over a field F is a curve of genus 1 defined over F, together with a given F-rational point on the curve. When F has characteristic different from 2, we can always find an affine equation for E of the form

$$E : y^2 = f(x),$$

where $f(x)$ in $F[X]$ is a cubic equation with distinct roots. Assuming that F has characteristic different from 2 and 3, the Weierstrass equation for E takes the form (see [32])

$$y^2 = x^3 + Ax + B$$

with coefficients in F. The discriminant Δ of E is defined by

$$\Delta = -16(4A^3 + 27B^2)$$

and is a fundamental invariant associated to the elliptic curve. Another important invariant is the conductor of an elliptic curve, which has the same prime divisors as the discriminant. The interested reader is referred to [32] and [33] for details on the basic arithmetic theory of elliptic curves.

We denote by $E(F)$ the set of solutions of E over F together with the "point at infinity" [32]. This set then has an abelian group structure. When F is a number field (i.e. a finite extension of $\mathbb{Q}$), it is further a celebrated result of Mordell and Weil that $E(F)$ is a finitely generated abelian group. Thus we define an important arithmetic invariant, called the *algebraic rank* of E, as

$$g_{E/F} := \text{rank of } E(F).$$

For example, the curve E_1 over $\mathbb{Q}$ defined by $y^2 = x^3 - x$ has algebraic rank zero while the curve E_2 over $\mathbb{Q}$ given by $y^2 = x^3 - 17x$ has algebraic rank 2. The curve E_1 has discriminant 64, and conductor equal to 32, while for E_2 the discriminant is $2^6 \times 17^3$ and the conductor is $2^5 \times 17^2$. Cremona's tables [15] gives a list of elliptic curves of small conductor along with their basic arithmetic data.

The primary reason for an abiding interest in this invariant is the important conjecture of Birch and Swinnerton-Dyer, formulated in the 1960's, based on very strong numerical data. For simplicity, we assume that the curve E is defined over $\mathbb{Q}$. Then the *Hasse-Weil L*-function of E, denoted $L(E, s)$ is a function of the complex variable s and is a vast generalisation of the classical Riemann-zeta function. It is defined using the integers $a_p := 1 + p - \#E(\mathbb{F}_p)$ as p varies over the prime numbers; here $\#E(\mathbb{F}_p)$ denotes the number of points on the reduction modulo p of the elliptic curve with coordinates in $\mathbb{F}_p$, with p a prime of good reduction (see [30], [32] for more details on reduction of elliptic curves). It was classically known that it converges when the real part of s is strictly greater than $3/2$. Let Δ_E denote the minimal discriminant of a generalised Weierstrass equation for the curve E [32, Chap. VII]. The Euler product expression for $L(E, s)$ is given by

$$L(E, s) = \prod_{p \nmid \Delta_E} (1 - a_p p^{-s} + (p^{1-2s}))^{-1} \prod_{p \mid \Delta_E} (1 - a_q . q^{-s})^{-1};$$

here for primes q dividing the discriminant of E, $a_q = 0, +1$ or -1 according as the singularity of the reduced elliptic curve over $\mathbb{F}_q$ is a node, or a cusp with rational or irrational tangents over $\mathbb{F}_q$. Further, it has a Dirichlet series expansion given by

$$L(E, s) = \sum_{n=1}^{\infty} a_n / n^s,$$

where the integers a_n are those defined above when n equals a prime p. The interested reader is referred to [32] and [15] for the explicit computation of the integers a_p.

The deep modularity results due to Wiles, Breuil *et al.* ([38], [4]) imply that the L-function has an analytic continuation for the entire complex plane. We remark that the *L*-function of *E* over a number field *F*, denoted $L(E/F, s)$ may be defined more generally for elliptic curves E/F, and it too is conjectured to have an analytic continuation over the entire complex plane. An elliptic curve $E/\mathbb{Q}$ is said to have *complex multiplication* if the endomorphism ring of *E* over an algebraic closure $\bar{\mathbb{Q}}$ of $\mathbb{Q}$ is strictly larger than the ring of integers. Both the curves E_1 and E_2 considered above are elliptic curves with complex multiplication as their endomorphism rings are given by the ring $\mathbb{Z}[i]$ of Gaussian integers. The element *i* acts as an endomorphism of the elliptic curve by sending a point (x, y) on the curve to $(-x, iy)$. At present the analytic continuation of the *L*-function is only known for elliptic curves with complex multiplication, thanks to work of Deuring and Weil. The *analytic rank* of *E*, denoted $r_{E/F}$ is defined to be the order of vanishing of $L(E/F, s)$ at $s = 1$. The Birch and Swinnerton-Dyer conjecture, in its weakest form, asserts that the analytic rank $r_{E/F}$ and the algebraic rank $g_{E/F}$ are equal.

Another important group associated to an elliptic curve defined over a number field *F* is the *Tate-Shafarevich group*, denoted $\text{III}(E/F)$. For any field *K*, and a discrete module *M* over the Galois group $G_K := \text{Gal}(\bar{K}/K)$, the first Galois cohomology group $H^1(G_K, M)$ is denoted by $H^1(K, M)$. For a place *v* of *F*, we denote the completion of *F* at *v* by F_v. The Tate-Shafarevich group of E/F is defined as the kernel

$$\text{III}(E/F) := \text{Ker}\left(H^1(F, E(\bar{F})) \longrightarrow \prod_v H^1(F_v, E(\bar{F}_v)) \right) \qquad (1)$$

of the natural restriction map, where the product on the right is taken over all places *v* of *F*. The Tate-Shafarevich group is analogous to the class group occurring in algebraic number theory (see §3). This group has an interesting geometric description in that it describes the defect of the 'local-global principle' for cubic curves. Thus, the non-trivial elements in it are classified by isomorphism classes of curves *X* defined over *F* which have the property that *X* becomes isomorphic to *E* over an algebraic closure $\bar{F}$ of *F* and $X(F) = \emptyset$ while $X(F_v) \neq \emptyset$ for all the completions. The Tate-Shafarevich group is one of the most mysterious groups occurring in arithmetic and is always conjectured to be finite. The exact formulae of the Birch and Swinnerton-Dyer conjecture even predicts its order, and surprisingly predicts that this order is usually, but not always, one.

The above discussion places elliptic curves at the heart of one of the deepest conjectures in modern number theory. We now turn to an ancient problem in number theory which has been illuminated by the conjecture of Birch and Swinnerton-Dyer. An integer $N \geq 1$ is said to be a *congruent number* if N is the area of a right angled triangle all of whose sides have *rational* length (see [5] for an excellent survey on this subject). The study of congruent numbers is over a thousand years old and a list of examples of congruent numbers occurs in Arab manuscripts from the 10th century A.D. A later folklore conjecture asserts that

Any positive integer $N \equiv 5,\ 6,\ 7 \mod 8$ is a congruent number. (2)

This conjecture turns out to be closely related to the study of elliptic curves. Specifically, it is easily seen that an integer N is congruent if and only if the elliptic curve

$$E_N : y^2 = x^3 - N^2 x \qquad (3)$$

defined over $\mathbb{Q}$ has the property that $E_N(\mathbb{Q})$ is infinite, or equivalently, $g_{E_N/\mathbb{Q}} > 0$. If one accepts the Birch and Swinnerton-Dyer conjecture, then it means that the analytic rank $r_{E/\mathbb{Q}} > 0$, in other words, that the L-function $L(E_N, s)$ vanishes at $s = 1$. The theory of root numbers ([3], see also §6), shows that in fact $L(E_N, s)$ always has a zero of odd multiplicity precisely for the integers N congruent to 5, 6, 7 modulo 8.

3. Iwasawa theory

Iwasawa theory is a relatively new area, owing its origins to the work of Iwasawa on cyclotomic $\mathbb{Z}_p$-extensions from the 1960's (see [20]). Henceforth, p will denote an odd prime. For a number field F, recall that the *class group* of F is the group of fractional ideals modulo the principal ideals, and is well-known to finite. The order of the class group is called the *class number* of F (cf. [26]). For s a complex variable, recall that $\zeta(s)$ is the classical Riemann-zeta function $\zeta(s) = \sum_{n=1}^{\infty} 1/n^s$. Let μ_p denote the group of p-th roots of unity. The philosophy emerging from Iwasawa's work initially provided an explanation for the link between special values of the Riemann zeta function and the class numbers of $\mathbb{Q}(\mu_p)$, as stated by Kummer's criterion (cf. [36]) below:

Theorem 3.1. *(Kummer's criterion) Let $K = \mathbb{Q}(\mu_p)$ and let h_K denote the class number of K. Then p divides h_K if and only if p divides the numerator of at least one of the values of $\zeta(-1), \zeta(-2), \ldots, \zeta(4 - p)$.*

Coates and Wiles recognised that techniques from Iwasawa theory could be used to attack the Birch and Swinnerton-Dyer conjecture. Recall that an elliptic curve $E/\mathbb{Q}$ is said to have *complex multiplication* if $\mathbb{Z} \subsetneq \mathrm{End}_{\bar{\mathbb{Q}}}(E)$ [32]. Coates and Wiles proved the first major general result about the Birch and Swinnerton-Dyer conjecture in [14] for elliptic curves with complex multiplication. A special case of their result is the following:

Theorem 3.2. *(Coates-Wiles) [14] Let $E/\mathbb{Q}$ be an elliptic curve with complex multiplication. Then $L(E,1) = 0$ whenever $E(\mathbb{Q})$ is infinite. In other words, $g_{E/\mathbb{Q}} > 0$ implies that $r_{E/\mathbb{Q}} > 0$.*

At present, Iwasawa theory has emerged as a systematic tool to attack the Birch and Swinnerton-Dyer conjecture using p-adic techniques. Let E be an elliptic curve over $\mathbb{Q}$, and let $\mathbb{Z}_p$ (resp. $\mathbb{Q}_p$) denote the ring of p-adic integers (resp. the field of p-adic numbers). In the complex world, no general connection between the behaviour of the complex L-function $L(E,s)$ at $s = 1$, and $E(\mathbb{Q})$ or $\mathrm{III}(E/\mathbb{Q})$ has ever been proven (there are some deep results due to Gross-Zagier-Kolyvagin, but they only apply to curves for which $L(E,s)$ has a zero at $s = 1$ of order at most 1). In the p-adic world however, such a link can be derived from the so-called "main conjectures" of Iwasawa theory, provided one replaces the complex L-function $L(E,s)$ by one of its p-adic avatars, at least when E has good ordinary reduction at the prime p. In particular, these main conjectures show that certain p-adic L-functions attached to E do have a zero at the point $s = 1$ in $\mathbb{Z}_p$ of order at least the rank of $E(\mathbb{Q})$ plus the number of copies of $\mathbb{Q}_p/\mathbb{Z}_p$ occurring in the p-primary subgroup $\mathrm{III}(E/\mathbb{Q})(p)$ of $\mathrm{III}(E/\mathbb{Q})$. We stress again that no result of this kind has ever been proven for the complex L-function. Here is an example of the type of result one can prove using these techniques:

Theorem 3.3. *(Coates, Liang, Sujatha) [10] Let $E/\mathbb{Q}$ be an elliptic curve with complex multiplication. For all sufficiently large good ordinary primes p, the number of copies of $\mathbb{Q}_p/\mathbb{Z}_p$ occurring in $\mathrm{III}(E/\mathbb{Q})(p)$ is at most $2p - g_{E/\mathbb{Q}}$.*

The basic idea of Iwasawa theory is to seek a simple connection between special values of L-functions and arithmetic of elliptic curves over certain infinite Galois extensions F_∞ of $\mathbb{Q}$. Viewed from this perspective, the Birch and Swinnerton-Dyer conjecture seems natural, as it elucidates how points of infinite order on E give rise to zeros of multiplicity at least $g_{E/\mathbb{Q}}$ of a p-adic L-function. Of course, it is beyond the scope of this article to develop the theory of p-adic L-functions in full detail and with greater precision. These are vast generalisations of the p-adic zeta functions that were studied by Kubota, Leopoldt and Iwasawa. We refer the interested reader to [6] for a detailed introduction to p-adic L-functions.

In the remaining sections, we shall outline how Iwasawa theory brings together three different strands viz. the conjecture (2) on congruent numbers, special values of L-functions, and algebraic questions on modules over Iwasawa algebras associated to compact p-adic Lie groups, with elliptic curves occurring as a common motif.

For an introduction to the Iwasawa theory of elliptic curves with complex multiplication, see [7]. The simplest elliptic curves without complex multiplication are the three curves of conductor 11 (see [15]). For a detailed study of their Iwasawa theory over the abelian extension $\mathbb{Q}(\mu_{p^\infty})$, see [11].

4. Iwasawa algebras

Let G be a profinite group, and p be an odd prime. The *Iwasawa algebra* of G, denoted $\Lambda(G)$ or $\mathbb{Z}_p[[G]]$, is the completed group algebra

$$\Lambda(G) := \varprojlim \mathbb{Z}_p[G/U];$$

here U varies over the open normal subgroups of G, $\mathbb{Z}_p[G/U]$ is the ordinary group ring over the finite group G/U and the inverse limit is taken with respect to the natural maps. Of special interest to us is the case when G is a compact p-adic Lie group. These groups were systematically studied by Lazard in his seminal work [24]. The simplest example is when G is isomorphic to $\mathbb{Z}_p$, in which case $\Lambda(G)$ is (non-canonically) isomorphic to the power series ring $\mathbb{Z}_p[[T]]$ in one variable. In classical cyclotomic theory, as considered by Iwasawa, one works with modules over this algebra (see [20], [12]). More generally, if $G \simeq \mathbb{Z}_p^d$, then $\Lambda(G)$ is isomorphic to the power series ring $\mathbb{Z}_p[[T_1, \cdots, T_d]]$ in d variables. If G is commutative, then $\Lambda(G)$ is a commutative $\mathbb{Z}_p$-algebra.

In classical Iwasawa theory, the infinite Galois extensions F_∞ that were considered were mostly abelian with Galois group G a commutative p-adic Lie group. As a specific example, consider the extension $F_\infty = F(\mu_{p^\infty})$ in which case G is open in $\mathbb{Z}_p$, and isomorphic to $\mathbb{Z}_p^\times$ if $F = \mathbb{Q}$. The Iwasawa algebra $\Lambda(G)$ is isomorphic to $\mathbb{Z}_p[\Delta][[T]]$, where Δ is cyclic of order dividing $p - 1$. Suppose E/F is an elliptic curve. For an odd prime p, consider the Galois extension

$$F_\infty = F(E_{p^\infty}) = \bigcup_{n \geq 1} F(E_{p^n}(\bar{F})), \qquad (4)$$

of F obtained by adjoining the coordinates of all the p-power division points of E. The module E_{p^∞} has a natural action of the Galois group $G(\bar{F}/F)$. If E has complex multiplication defined over F, i.e. $\text{End}_F(E) \neq \mathbb{Z}$, then G is abelian and contains an open subgroup isomorphic to $\mathbb{Z}_p^2$.

It is important to consider elliptic curves over $\mathbb{Q}$ (or more generally over a number field) without complex multiplication. Indeed, elliptic curves with complex multiplication are rather special and those without complex multiplication are more abundant. For such curves, the elements in the endomorphism ring correspond to multiplication by an integer n, given by the group law and hence the endomorphism ring is isomorphic to the ring of integers. It is a deep result of Serre [31] that the extension (4) is a non-commutative p-adic Lie extension. In fact, Serre also proved that the Galois group $G := G(F_\infty/F)$ for such elliptic curves is an open subgroup of $GL_2(\mathbb{Z}_p)$ and is equal to it for almost all primes p. The Iwasawa algebra $\Lambda(G)$ is thus highly non-commutative. Another natural example of a non-commutative p-adic Lie extension is given by the so-called "False Tate extension", obtained by adjoining all p-power roots of unity and the p-power roots of an integer m which is p-power free, i.e.

$$ F_\infty := F(\mu_{p^\infty}, m^{1/p^\infty}). \tag{5} $$

In this case the Galois group G is an open subgroup of the semi-direct product $\mathbb{Z}_p^\times \ltimes \mathbb{Z}_p$.

Lazard proved that for any compact p-adic Lie group G, the Iwasawa algebra is a left and right noetherian ring. Further, if G is pro-p and has no elements of order p, then $\Lambda(G)$ is a local domain, in the sense that it has no zero divisors, and the set of non-units form a (unique) two-sided maximal ideal. In particular for $G = \text{Gal}(F_\infty/F)$ with F_∞ as in (4), the Iwasawa algebra $\Lambda(G)$ is a left and right noetherian, local domain whenever $p \geq 5$. In the last decade, these algebras have been investigated more thoroughly (see [35]). The analogue in the non-commutative setting, of classical regular local rings in commutative algebra, is that of 'Auslander-regular' rings (see [35]) and in all the cases of non-commutative p-adic Lie extensions mentioned above, the corresponding rings are Auslander regular. More precisely, we have:

Theorem 4.1. *(Venjakob) [35] Let G be a compact p-adic analytic group without p-torsion, and of dimension d when considered as an analytic manifold. Then $\Lambda(G)$ is an Auslander regular local domain of injective dimension equal to d.*

The main advantage in having this nice extra structure on $\Lambda(G)$ is that it provides a 'dimension theory' on the category of finitely generated modules over $\Lambda(G)$. The dimension of $\Lambda(G)$ itself is $d + 1$. This in turn, affords the definition of *pseudonull* modules. Asssume that G is as in Theorem 4.1, and let M be a finitely generated module over $\Lambda(G)$. Then M is pseudonull if the dimension of M is less than or equal to $d - 1$. There is an equivalent characterisation of pseudonull modules using homological algebra (see [35]), and it coincides with the classical notion of pseudonull modules in commutative algebra. Note that in the simple case when $\Lambda(G) \simeq \mathbb{Z}_p[[T]]$, pseudonull modules are precisely the finite modules.

In the commutative case, there is also a well-known classical structure theorem for finitely generated modules over $\Lambda(G)$, due to Iwasawa and Serre (see [2], [12, Appendix]). We say that two finitely generated modules M and N over $\Lambda(G)$ are pseudoisomorphic if there is a $\Lambda(G)$ homomorphism between them whose kernel and cokernel are pseudonull.

Theorem 4.2. *Suppose G is a commutative p-adic Lie group with no elements of order p and let $\Lambda(G)$ be its Iwasawa algebra. Let M be a finitely generated torsion module over $\Lambda(G)$. Then there is a pseudoisomorphism*

$$M \rightarrow \overset{k}{\underset{i=1}{\oplus}} \Lambda(G)/\mathfrak{p}_i^{n_i}$$

where the $\mathfrak{p}_i$ are prime ideas of height one and n_i are positive integers.

It is well-known [2] that the prime ideals of height one in $\Lambda(G)$, for G as in the theorem above, are principal. Let I denote the ideal defined as the product $I := \prod_i \mathfrak{p}_i^{n_i}$. It is called the *characteristic ideal* of M (this is well-defined, see [2]) and denoted by char_M. A generator of the characteristic ideal is the *characteristic power series* of M and is well-defined up to a unit in the Iwasawa algebra. The characteristic power series plays a central role in the formulation of the main conjecture, and will be discussed in the next section.

In the non-commutative case, the fact that the Iwasawa algebra $\Lambda(G)$ is Auslander regular can be exploited to prove a more rudimentary structure theorem. A module over $\Lambda(G)$ will be assumed to be a left module. A finitely generated $\Lambda(G)$-module M is said to be *torsion*, if every element of M is annihilated by a non-zero divisor of $\Lambda(G)$. A module M is said to be *reflexive* if the natural map $M \rightarrow M^{++}$ is an isomorphism; here M^+ denotes the dual module $\mathrm{Hom}_{\Lambda(G)}(M, \Lambda(G))$.

Theorem 4.3. *(Coates-Schneider-Sujatha) [13] Suppose G is a compact p-adic analytic Lie group of dimension d with no element of order p. Let M be a finitely generated torsion module over $\Lambda(G)$. Then there is a homomorphism*

$$f : M \to \bigoplus_{i=1}^{n} \Lambda(G)/J_i$$

where the J_i are reflexive ideals and f has pseudonull kernel and cokernel.

5. Main conjectures

The aim of this section is to outline the philosophy and the fomulation of the "main conjectures" in Iwasawa theory for elliptic curves. We do not even pretend to attempt a discussion of the steps involved in the formulation of these conjectures in full detail. Our goal shall be largely confined to giving the reader a flavour of what goes under the rubric of main conjectures. The basic idea is to first attach an algebraic invariant and an analytic invariant to certain canonically defined arithmetic modules over the Iwasawa algebra $\Lambda(G)$ of the Galois group G of an infinite p-adic Lie extension F_∞. The analytic invariant has the property of interpolating special values of the complex L-function, with the interpolation formula being explicit. The main conjecture asserts the equality of these invariants. We discuss a few concrete examples below.

Iwasawa in his classic study of $\mathbb{Z}_p$-extensions [20] studied the growth of ideal class groups in the cyclotomic $\mathbb{Z}_p$-extensions, and was the first to formulate the main conjecture for the field $\mathbb{Q}(\mu_{p^\infty})$. Here is a brief explanation of one version of his main conjecture. He related the arithmetic of the "Tate motive" over the extension $F_\infty = \mathbb{Q}(\mu_{p^\infty})^+$ (here $+$ denotes the maximal real subfield of $\mathbb{Q}(\mu_{p^\infty})$) to special values of the Riemann-zeta function, via the Kubota-Leopoldt p-adic zeta function. This element ζ_p is viewed as a pseudo-measure on the p-adic Lie group $G = \mathrm{Gal}(F_\infty/\mathbb{Q})$, and also as belonging to an explicit localisation of the Iwasawa algebra $\Lambda(G)$. It has the following interpolation property, where χ is the cyclotomic character giving the action of Galois on μ_{p^∞}:-

$$\int_G \chi(g)^k \, d\zeta_p = (1 - p^{k-1})\zeta(1-k)$$

for all even integers $k \geq 2$. The corresponding arithmetic module is as follows. Let X_∞ denote the maximal abelian extension of F_∞ that is unramified outside p. Then X_∞ has a natural structure of a finitely generated $\Lambda(G)$-module and it is a deep result of Iwasawa that is is a torsion module over $\Lambda(G)$. The main conjecture asserts that the characteristic ideal of X_∞ is equal to the ideal $\zeta_p.I_G$, where I_G is the augmentation ideal of $\Lambda(G)$, i.e. the kernel of the natural quotient map $\Lambda(G) \to \mathbb{Z}_p$. Iwasawa himself proved a remarkable general theorem about the arithmetic of the field F_∞, involving a module formed out of cyclotomic units, which is closely related to X_∞. In particular, this theorem implies his main conjecture when the class number of $\mathbb{Q}(\mu_p)^+$ is prime to p. The first unconditional proof of the main conjecture was given by Mazur-Wiles [25] and Wiles gave a second proof in [39], beautifully extending ideas of Ribet. A simpler proof using Iwasawa's original approach, along with work of Thaine, Kolyvagin and Rubin on Euler systems [34], [23], [28], is given in [12].

To formulate these main conjectures for elliptic curves, one studies the arithmetic of E over infinite p-adic Lie extensions of a number field F. The p-adic L-functions then seem to mysteriously arise from some natural G-modules describing the arithmetic of E over these p-adic Lie extensions. We shall sketch the formulation of the main conjecture in the important case of elliptic curves with complex multiplication, which was first considered by Coates-Wiles. Of course, the general case of elliptic curves without complex multiplication lies much deeper and is more technical. Let $E/\mathbb{Q}$ be an elliptic curve with complex multiplication by the ring of integers $\mathcal{O}_K$ of an imaginary quadratic field K of class number one. Suppose p is a prime such that p splits as $p = \mathfrak{p}\mathfrak{p}^*$ in $\mathcal{O}_K$, and assume that E has good ordinary reduction at $\mathfrak{p}$ and $\mathfrak{p}^*$. By the classical theory of complex multiplication due to Deuring and Weil, it is well-known that the complex L-function $L(E/\mathbb{Q}, s)$ is the Hecke L-function $L(\psi_E, s)$ where ψ_E is a certain Grössencharacter (see [32]).

The p-adic Lie extension that we consider is the extension

$$F_\infty = \bigcup_{n \geq 1} K(E_\mathfrak{p}^n) \tag{6}$$

obtained by adjoining all the $\mathfrak{p}$-division points of the elliptic curve to K. The Galois group of F_∞ over K is isomorphic to $\mathbb{Z}_p^\times$ and we denote by K_∞ the unique $\mathbb{Z}_p$-extension contained in F_∞. Let $\Gamma = \mathrm{Gal}(K_\infty/K)$, then the Iwasawa algebra $\Lambda(\Gamma)$ is isomorphic to $\mathbb{Z}_p[[T]]$ (see §4). The p-adic L-function is then an element $H_\mathfrak{p}(T)$ in $\mathcal{I}[[T]]$, where $\mathcal{I}$ denotes the ring of integers in the completion of the maximal unramified extension of $\mathbb{Q}_p$. It interpolates the values of the complex L-function in that we have

$$\Omega_\mathfrak{p}^{-n} H_\mathfrak{p}((1+p)^n - 1) = \Omega_\infty^{-n}(n-1)! L(\bar{\psi}_E^n, n) \left(1 - \frac{\psi_E^n(\mathfrak{p})}{N\mathfrak{p}}\right),$$

for appropriate complex and p-adic periods Ω_∞ and $\Omega_\mathfrak{p}$ respectively, of the elliptic curve (see [7] for a detailed exposition).

Classical descent theory [32] already points to the arithmetic module that one should consider. This is the *Selmer group* which we define below. Let $\mathcal{M}$ be any Galois extension of a number field F. For each non-archimedean place w of $\mathcal{M}$, let $\mathcal{M}_w$ be the union of the completions at u of all finite extensions of F contained in $\mathcal{M}$. The p^∞-Selmer group of E over $\mathcal{M}$ is defined by

$\mathrm{Sel}_p(E/\mathcal{M})$

$$= \mathrm{Ker}\left(H^1(\mathrm{Gal}(\bar{\mathcal{M}}/\mathcal{M}), E_{p^\infty}) \to \prod_w H^1(\mathrm{Gal}(\bar{\mathcal{M}}_w/\mathcal{M}_w), E(\bar{\mathcal{M}}_w)) \right), \quad (7)$$

where w runs over all non-archimedean places of $\mathcal{M}$, and the map is given by natural restriction. The Galois group of $\mathcal{M}$ over F operates on $\mathrm{Sel}_p(E/\mathcal{M})$ and we have an exact sequence

$$0 \to E(\mathcal{M}) \otimes_{\mathbb{Z}_p} \mathbb{Q}_p/\mathbb{Z}_p \to \mathrm{Sel}_p(E/\mathcal{M}) \to \mathrm{III}(E/\mathcal{M})(p) \to 0. \quad (8)$$

Here $\mathrm{III}(E/\mathcal{M})$ denotes the Tate-Shafarevich group of E over $\mathcal{M}$, which is the inductive limit of $\mathrm{III}(E/L)$ as L varies over all finite extensions of F in $\mathcal{M}$, and for any abelian group A, $A(p)$ is the submodule consisting of all elements annihilated by a power of p. We shall consider the Pontryagin dual

$$X_p(E/\mathcal{M}) = \mathrm{Hom}(\mathrm{Sel}_p(E/\mathcal{M}, \mathbb{Q}_p/\mathbb{Z}_p). \quad (9)$$

which is a compact module over the Galois group $\mathrm{Gal}(\mathcal{M}/K)$. The dual Selmer group considered as a module over the Iwasawa algebra, simultaneously reflects both the arithmetic of the elliptic curve and the special values of the complex L-function. Further, by virtue of the additional Galois module structure, it encodes information about $E(L)$ and $\mathrm{III}(E/L)$ for all finite extensions L of F in $\mathcal{M}$.

Suppose now that $E/\mathbb{Q}$ is an elliptic curve with complex multiplication such that $\mathrm{End}_K(E) \simeq \mathcal{O}_K$, and let K_∞ be the $\mathbb{Z}_p$-extension of K contained in F_∞ (cf. (6)). The Selmer group $\mathrm{Sel}_\mathfrak{p}(E/K_\infty)$ is similarly defined as the kernel of the restriction map

$$\mathrm{Ker}\left(H^1\,\mathrm{Gal}(\bar{\mathbb{Q}}/K_\infty), E_{\mathfrak{p}^\infty}) \to \prod_w H^1(\mathrm{Gal}(\bar{K}_{\infty,w}/K_{\infty,w}), E(\bar{K}_{\infty,w})) \right)$$

where w runs over all the non-archimedean places of K_∞. Clearly the Galois group of K_∞ over K operates on $\mathrm{Sel}_\mathfrak{p}(E/K_\infty)$ and we have an exact sequence

$$0 \to E(K_\infty) \otimes_{\mathcal{O}_K} (K_\mathfrak{p}/\mathcal{O}_\mathfrak{p}) \to \mathrm{Sel}_\mathfrak{p}(E/K_\infty) \to \mathrm{III}(E/K_\infty)(\mathfrak{p}) \to 0.$$

Here $\mathrm{III}(E/K_\infty)$ denotes the Tate-Shafarevich group of E over K_∞ and for any $\mathcal{O}_K$-module A, $A(\mathfrak{p})$ denotes the submodule consisting of elements annihilated by some power of a generator of $\mathfrak{p}$. As before (cf. (9)), we consider the compact dual, which we denote by $X_\mathfrak{p}(E/K_\infty)$. This is a finitely generated module over $\Lambda(\Gamma)$, which is torsion, thanks to a result of Coates-Wiles [14]. By the structure theorem described in §4, we can define the characteristic power series of the dual Selmer group, which we denote by $B_\mathfrak{p}(T) \in \mathbb{Z}_p[[T]]$. The one variable main conjecture, proved by Rubin [29], is the following deep result:

Theorem 5.1. *(One variable main conjecture) [29] We have*

$$H_\mathfrak{p}((1+p)(1+T) - 1)\mathcal{I}[[T]] = B_\mathfrak{p}(T)\mathcal{I}[[T]].$$

Let $E/\mathbb{Q}$ be an elliptic curve without complex multiplication, and let p be a prime of good ordinary reduction. In this case, the formulation of the main conjecture over the cyclotomic $\mathbb{Z}_p$-extension can be found in [18]. For the algebraic invariant, a deep result of Kato [22] proves that the dual Selmer group is a finitely generated torsion module over the corresponding Iwasawa algebra, and hence the characteristic ideal can be defined as before. Moreover, Kato proves that the p-adic L-function is divisible by this characteristic ideal. Completing the proof of the main conjecture is however considerably harder, and to date, a full proof has not been published, (Skinner and Urban have announced results in this direction).

For nonabelian p-adic Lie extensions as in the division field extension (4) or the false Tate extension (5), even the precise formulation of the main conjecture is far from obvious. Let G be the corresponding Galois group and $\Lambda(G)$ the associated Iwasawa algebra. Though the dual Selmer group is known to be finitely generated as a (left) module over the corresponding Iwasawa algebra $\Lambda(G)$, and is even conjectured to be torsion (in fact, there is even a stronger conjecture, see [9]), there is no well-defined analogue of the characteristic ideal. A main conjecture in this set-up is formulated in [9], and the principal novelty in these non-commutative examples is the use of algebraic K-theory [1]. The algebraic and analytic invariants are elements of the group $K_1(R)$, where R is an explicit localisation of the Iwasawa algebra $\Lambda(G)$. The existence of a canonical Ore set in $\Lambda(G)$ makes this explicit localisation possible. Furthermore, this formulation can be intrinsically linked to Iwasawa theory of the elliptic curve over the cyclotomic extension, which is a quotient of F_∞ in the examples considered above. For a commutative ring R, $K_1(R)$ may be identified with the units in R and therefore, the occurrence of K_1 in the non-commutative set-up may be viewed as a natural extension of the commutative context. The main conjecture then predicts the equality of the analytic and algebraic invariants, as elements in the K-group. We do not go into any further details but state that the non-commutative phenomenon is vastly different in one other aspect. Namely, it has infinite families of self-dual Artin representations of G (these are representations that factor through a finite quotient of G) and thus gives rise to twists of complex L-functions. The interpolation property of the p-adic L-function then has to take into account these twisted L-values, in the formulation of the main conjecture. This in turn leads to interesting connections with root numbers, which we shall touch upon in the next section. When E has supersingular reduction at p [32], we still have no idea how to formulate a non-commutative main conjecture.

6. Applications and examples

In this final section, we mention a few theorems that are proved using Iwasawa theory. We remark that even though the main conjecture has only been established in a few cases, it provides great insights into the Birch and Swinnerton-Dyer conjecture. Kakde [K] has recently proven the existence of the p-adic L-function and made important progress towards the main conjecture in the non-commutative case for the Tate motive over p-adic Lie extensions of totally real number fields. Another interesting phenomenon is the connection between root numbers and non-commutative Iwasawa theory which is studied in [8]. In particular, these results give information on the growth of the Mordell-Weil ranks along finite layers of the false Tate extension and the division field extension. We first recall the definition of the root number.

Let $E/\mathbb{Q}$ be an elliptic curve. The modified L-function denoted $\Lambda(E, s)$, s a complex variable, is defined by

$$\Lambda(E, s) = (2\pi)^{-s}\Gamma(s)L(E, s).$$

By the modularity result of Wiles *et al.*, this function is entire and satisfies the functional equation

$$\Lambda(E, s) = \omega_E N_E^{1-s}\Lambda(E, 2 - s),$$

where $\omega_E = \pm 1$ is the *root number* and N_E is the conductor of E [32]. We have

$$\omega_E = (-1)^{r_{E/\mathbb{Q}}} \tag{10}$$

where $r_{E/\mathbb{Q}}$ is the analytic rank of E. Root numbers can also be defined over finite extensions of $\mathbb{Q}$. The study of root numbers by Rohrlich [27] along the cyclotomic extension, combined with the deep result of Kato that the dual Selmer group of E is torsion over the Iwasawa algebra [22] yields the following result:

Theorem 6.1. *[22, 27] For every prime p, $E(\mathbb{Q}(\mu_{p^\infty}))$ is a finitely generated abelian group.*

We next consider a false Tate extension tower. Fix an integer $m > 1$, which is assumed to be p-power free. Define

$$L_n = \mathbb{Q}(m^{1/p^n}), \ K_n = \mathbb{Q}(\mu_{p^n}), \ F_n = \mathbb{Q}(\mu_{p^n}, m^{1/p^n}),$$

and consider the false Tate extension

$$F_\infty = \bigcup_{n \geq 0} F_n$$

with Galois group G. Let H be the normal subgroup

$$H := \mathrm{Gal}(F_\infty/\mathbb{Q}(\mu_{p^\infty})) \simeq \mathbb{Z}_p.$$

Then G is isomorphic to the semi-direct product of $\mathbb{Z}_p^\times$ and $\mathbb{Z}_p$. The extensions L_n are not Galois, while F_n are nonabelian Galois extensions, and the Artin representations of G can be fully described. Put

$$Y(E/F_\infty) = X_p(E/F_\infty)/X_p(E/F_\infty)(p),$$

where $X_p(E/F_\infty)(p)$ is the p-primary submodule of the dual Selmer group (9). For any finite extension M of $\mathbb{Q}$, we define

$$s_{E/M,p} = \mathbb{Z}_p - \text{corank of the Selmer group of } E \text{ over } M. \tag{11}$$

The study of root numbers, combined with results from Iwasawa theory, yields the following theorem:

Theorem 6.2. *[8, Theorem 4.8] Assume that E has good ordinary reduction at p and that $Y(E/F_\infty)$ is finitely generated as a $\Lambda(H)$-module, with $\Lambda(H)$-rank 1. Then for all $n \geq 1$, we have*

$$s_{E/L_n,p} = n + s_{E/\mathbb{Q},p}, \quad s_{E/F_n,p} = p^n - 1 + s_{E/K_1,p}.$$

As a specific numerical example where the above theory can be applied, we consider the elliptic curve $E/\mathbb{Q}$ of conductor 11 defined by

$$E : y^2 + y = x^3 - x^2, \tag{12}$$

and the prime $p = 7$.

Theorem 6.3. *Let F_∞ be a false Tate extension. For the elliptic curve E as in (12), and $p = 7$, we have the algebraic rank*

$$g_{E/L_n} \geq n, \quad (n = 1, 2, 3 \cdots)$$

provided $\mathrm{III}(E/L_n)(7)$ is finite.

We remark that even for $n = 1$, it is numerically very difficult to find points of infinite order in $E(L_1)$. Surprisingly, Iwasawa theory also provides lower bounds in some cases.

Theorem 6.4. *Assume that m is any 7-power free integer with prime factors in the set $\{2, 3, 7\}$. Then for E as in (12), and all integers $n = 2, 3, \ldots$, we have*

$$g_{E/L_n} \leq n$$

with equality if and only if $\mathrm{III}(E/L_n)(7)$ is finite.

A natural question that arises in light of (10) and the Birch and Swinnerton-Dyer conjecture is whether the root number and the algebraic rank have the same parity. Assuming that the Tate-Shafarevich group is finite, this is equivalent to the question whether $s_{E/\mathbb{Q},p}$ (cf. (8), (11)) and the root number have the same parity. An important general result in this direction has been proved by T. Dokchitser and V. Dokchitser [16]:

Theorem 6.5. *(T. Dokchitser and V. Dokchitser [16]) Let $E/\mathbb{Q}$ be an elliptic curve. Then for any prime p, the root number ω_E and $s_{E/\mathbb{Q},p}$ have the same parity.*

We end this article by showing how these results enable us to go considerably closer to proving the folklore conjecture (2) on congruent numbers.

Theorem 6.6. *Assume $N \equiv 5, 6, 7 \bmod 8$, and let E_N be the elliptic curve defined by (3). If the p-primary torsion part $\mathrm{III}(E_N/\mathbb{Q})(p)$ is finite for some prime p, then N is congruent.*

Proof. As remarked earlier, it is known from the theory of L-functions that $L(E_N, s)$ vanishes to odd order at $s = 1$ for N as in the theorem. By the parity theorem 6.5, we therefore see that $s_{E_N/\mathbb{Q},p}$ is odd for all primes p. Suppose there exists a prime p such that $\text{III}(E_N/\mathbb{Q})(p)$ is finite. Then by the exact sequence (8), we have $g_{E_N/\mathbb{Q}} \geq 1$ and hence E_N has a point of infinite order. By our remarks at the end of §2, this implies that N is a congruent number. $\qquad\qquad\square$

References

1. H. BASS, Algebraic K-theory, W. A. Benjamin, Inc., New York-Amsterdam (1968).
2. N. BOURBAKI, Elements of Mathematics, Commutative Algebra, Chapters 1-7, Springer (1989).
3. B. BIRCH, G. STEVENS, *The parity of the rank of the Mordell-Weil group*,Topology **5** (1966), 295–299.
4. C. BREUIL, B. CONRAD, F. DIAMOND, R. TAYLOR, *On the modularity of elliptic curves over* **Q***: wild 3-adic exercises*, J. Amer. Math. Soc. **14** (2001), 843–939.
5. J. COATES, *Congruent number problem*, Q. J. Pure Appl. Math. **1** (2005), 14–27.
6. J. COATES, *p-adic L-functions and Iwasawa's theory*, in Algebraic Number fields (Durham Symposium); ed. A. Frohlich, Academic Press (1977), 269–353.
7. J. COATES, *Elliptic curves with complex multiplication and Iwasawa theory*, Bull. LMS. **23** (1991), 321–350.
8. J. COATES, T. FUKAYA, K. KATO, R. SUJATHA, *Root numbers, Selmer groups and non-commutative Iwasawa theory*, Jour. Alg. Geom. (To appear).
9. J. COATES, T. FUKAYA, K. KATO, R. SUJATHA, O. VENJAKOB, *The GL_2 main conjecture for elliptic curves without complex multiplication*, Publ. Math. IHES **101** (2005), 163–208.
10. J. COATES, Z. LIANG, R. SUJATHA, Tate Shafarevich groups of elliptic curves with comples multiplication, arXiv:0901.3832v1 [math.NT], To appear in Journal of Algebra.
11. J.COATES, R. SUJATHA, Galois cohomology of elliptic curves, Tata Institute of Fundamental Research Lectures on Mathematics **88**, Narosa Publishing House, New Delhi, (2000).
12. J. COATES, R. SUJATHA, Cyclotomic fields and zeta values, Springer Monographs in Mathematics, Springer (2006).
13. J. COATES, P. SCHNEIDER, R. SUJATHA, *Modules over Iwasawa algebras*, J. Inst. Math. Jussieu **2** (2003), 73–108.
14. J. COATES, A. WILES, *On the conjecture of Birch and Swinnerton-Dyer*, Invent. Math. **39** (1977), 223–251.
15. J. CREMONA, Algorithms for modular elliptic curves, Second edition, Cambridge University Press, Cambridge, (1997).

16. T. DOKCHITSER, V. DOKCHITSER, *On the Birch-Swinnerton Dyer quotients modulo squares*, Ann. of Math. (To appear).

17. W. FULTON, Algebraic curves. An introduction to algebraic geometry, Notes written with the collaboration of Richard Weiss, Reprint of 1969 original,Advanced Book Classics. Addison-Wesley Publishing Company, Advanced Book Program, Redwood City, CA, 1989.

18. R. GREENBERG, *Iwasawa theory for elliptic curves*, in Arithmetic theory of elliptic curves (Cetraro, 1997), Lecture Notes in Math., 1716, Springer, Berlin, (1999), 51–144.

19. J. HOYRUP, *The Babylonian Cellar Text BM85200 + VAT 6599 Retranslation and Analysis*, in Amphora Festschrift for Hans Wussing on the occasion of his 65th birthday, ed. Hans Wussing, Sergei Sergeewich Demidov, Birkhäuser, 315–338.

20. K. IWASAWA, *On Z_l-extensions of algebraic number fields*, Ann. of Math. **98** (1973), 246–326.

21. M. KAKDE, *The Main Conjecture of non-commutative Iwasawa theory for totally real fields*, Jour. Alg. Geom. (To appear).

22. K. KATO, *p-adic Hodge theory and values of zeta functions of modular forms*,in Cohomologies *p*-adiques et applications arithmétiques. III. Astérisque **295** (2004), ix, 117–290.

23. V. KOLYVAGIN, *Euler systems*, in The Grothendieck Festschrift, Vol. II, Progr. Math. **87**, Birkhüser Boston, Boston, MA (1990), 435–483.

24. M. LAZARD, *Groupes analytiques p-adiques*, Publ. Math. IHES **26** (1965) 389–603.

25. B. MAZUR, A. WILES, *Class fields of abelian extensions of* $\mathbb{Q}$, Invent. Math. **76** (1984), 179–330.

26. J. NEUKIRCH, Algebraic number theory, Grundlehren der Mathematischen Wissenschaften **322** Springer-Verlag, Berlin (1999).

27. D. ROHRLICH, *On L-functions of elliptic curves and cyclotomic towers*, Invent. Math. **18** (1972), 183–266.

28. K. RUBIN, *The "main conjectures" of Iwasawa theory for imaginary quadratic fields*, Invent. Math. **103** (1991), 25–68.

29. K.RUBIN, *The one-variable main conjecture for elliptic curves with complex multiplication*, in *L*-functions and arithmetic (Durham, 1989), London Math. Soc. Lecture Note Ser., **153**, Cambridge Univ. Press, Cambridge, 1991, 353–371.

30. J.-P. SERRE, J. TATE, *Good reduction of abelian varieties*, Ann. of Math. **88** (1968), 492–517.

31. J.-P. SERRE, *Propriétés galoisiennes des points d'ordre fini des courbes elliptiques*, Invent. Math. **15** (1972), 259–331.

32. J. SILVERMAN, The arithmetic of elliptic curves,Corrected reprint of the 1986 original. Graduate Texts in Mathematics **106** Springer-Verlag, New York, (1992).

33. J. SILVERMAN, J. TATE, Rational points on elliptic curves, Undergraduate Texts in Mathematics, Springer-Verlag, New York, (1992).

34. F. THAINE, *On the ideal class groups of real abelian number fields*, Ann. of Math. **128**, (1988), 1–18.

35. O. VENJAKOB, *On the structire theory of Iwasawa algebra of a p-adic Lie group*,J. Eur. Math. Soc. **4** (2002), 271–311.

36. L. WASHINGTON, Introduction to Cyclotomic fields, Graduate Texts in Mathematics **83**, Springer (1982).

37. A. WEIL,Number theory. An approach through history. From Hammurapi to Legendre, Birkhuser Boston, Inc., Boston, MA, (1984).

38. A. WILES, *Modular elliptic curves and Fermat's last theorem*, Ann. of Math. **141** (1995), 443–551.

39. A. WILES, *The Iwasawa conjecture for totally real fields*, Ann. of Math. **131** (1990), 493–540.

PART B

Contributed Short Talks

TRICRITICAL POINTS AND LIQUID-SOLID CRITICAL LINES

ANNELI AITTA

Institute of Theoretical Geophysics
Department of Applied Mathematics and Theoretical Physics
University of Cambridge
Wilberforce Road
Cambridge, CB3 0WA, UK

Tricritical points separate continuous and discontinuous symmetry breaking transitions. They occur in a variety of physical systems and their mathematical models. A tricritical point is used to determine a liquid-solid phase transition line in the pressure-temperature plane [12]. Excellent experimental agreement has been obtained for iron, the material having the most high pressure data. This allows extrapolation to much higher pressures and temperatures than available experimentally. One can predict the temperature at the liquid-solid boundary in the Earth's core where the pressure is 329 GPa. Light matter, present as impurities in the core fluid, is found to generate about a 600 K reduction of this temperature.

Keywords: Tricritical point; phase transitions; critical phenomena; iron melting curve; temperature in the Earth's core.

1. Introduction

Melting or solidification is a first order phase transition since the order changes discontinuously from liquid to solid. Landau (1937) gave a theoretical description for first order phase transitions and the point where they change to second order phase transitions (with a continuous change of order) [1]. Such a point was later named a tricritical point by Griffiths (1970) [2]. Tricritical points occur in a variety of physical systems. Examples of tricritical points are presented in Table 1.1 with the corresponding adjustable variables. Experimentally they were first found in fluid mixtures, compressed single crystals and magnetic and ferroelectric systems (see old reviews in [3] and [4]). Paper [11] is an example of two-dimensional melting.

The rest of this paper provides bifurcation theoretical analysis for the solidification/melting problem for iron [12], following the earlier work in

Table 1.1.   Examples of tricritical points.

Physical system	Variable 1	Variable 2
3He-4He mixtures [2]	Density or concentration	Temperature
Vortex-lattice melting [5]	Magnetic field	Temperature
Liquid crystals [6]	Concentration	Temperature
Cold Fermi gas [7]	Spin polarization	Temperature
Ferroelectrics [8]	Pressure or electric field	Temperature
Metamagnets [9]	Pressure or magnetic field	Temperature
Structural transition [10]	Pressure	Temperature
Melting on graphite [11]	Coverage	Temperature
Solidification [12]	Pressure	Temperature
Taylor-Couette vortex pair [13]	Aspect ratio	Rotation rate

Paper [13] which presents the symmetry breaking analysis in the first experimentally studied nonequilibrium tricritical point.

2. Landau theory

Following Landau [1], an order parameter x can be used to describe first order phase transitions which change to be second order at a tricritical point. Here $x = 0$ for the more ordered solid phase which occurs at lower temperature, and in the less ordered liquid phase, $x \neq 0$. The Gibbs free energy density is proportional to the Landau potential, which needs to be a sixth order polynomial in x:

$$\Phi = x^6/6 + gx^4/4 + \varepsilon x^2/2 + \Phi_0. \tag{1}$$

A set of examples of $\Phi - \Phi_0$ is shown in Fig. 2.1. No higher order terms in x appear in Φ since they can be eliminated using coordinate transformations as in bifurcation theory [14]. This method also scales out any dependence on physical parameters of the coefficient of the $x6$ term. Generally Φ_0, ε and g depend on the physical parameters and for solidification they are pressure P and temperature T. In equilibrium, the order parameter takes a value where the potential Φ has a local or global minimum. The minima of Φ, the solutions of

$$x^5 + gx^3 + \varepsilon x = 0 \tag{2}$$

at which $d^2\Phi/dx^2$ is positive, give three stable equilibrium states provided $0 < \varepsilon < g^2/4$ and $g < 0$. They are at

$$x = 0 \text{ and } x = \pm\sqrt{-g/2 + \sqrt{g^2/4 - \varepsilon}}. \tag{3}$$

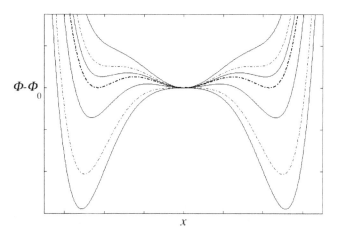

Fig. 2.1. Order parameter x dependent part of the Landau potential (1) for a fixed $g < 0$ and various values of ε.

The thermodynamic transition from liquid to solid occurs when all three minima of the potential are equally deep: $\Phi - \Phi_0 = 0$ at those three values of x. This happens when

$$\varepsilon = 3g^2/16 \tag{4}$$

corresponding to the middle dash-dotted curve in Fig. 2.1. The liquid phase is then in thermal equilibrium with the solid phase. If $\varepsilon > 3g^2/16$ the solid is preferred, and if $\varepsilon < 3g^2/16$ the liquid. There are also two other critical conditions: Liquid phase exists as an unfavoured state until the potential changes from having three minima to one minimum (the highest dash-dotted curve in Fig. 2.1), that is, at

$$\varepsilon = g^2/4. \tag{5}$$

The solid phase exists as an unfavoured state until $\varepsilon = 0$ where the potential changes from having three to two minima (the lowest dash-dotted curve in Fig. 2.1).

These liquid-solid phase transitions can be presented using simple bifurcation diagrams as in Fig. 2.2. In the direction where T increases (see Fig. 2.2a), for $T < T_S$ the solid state (having $x=0$) is the only possible state of the system. At higher temperatures solid state is preferred but liquid state is possible as an unfavoured stable state until $T = T_M$ where the melting occurs. At higher temperatures the liquid state is preferred but solid state can occur as an unfavoured stable state until $T = T_L$ beyond

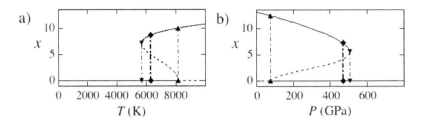

Fig. 2.2.   Bifurcation diagrams for non-negative orderparameter x as function of (a) temperature and (b) pressure. Symbols show the values of the critical conditions: downward triangle, diamond and triangle correspond to subindex s, m and l, respectively.

which only liquid exists because the solid state is unstable. Now consider the direction where P increases (see Fig. 2.2b which has $T > T_L$). For $P < P_L$ only liquid state is stable. There solid state is unstable and thus not a possible state of the system. At $P = P_L$ the unstable solid state bifurcates to a stable solid state and two unstable liquid states (only the branch with positive x is shown) which turn backward to be stable states at $P = P_S$. For $P_L < P < P_M$ liquid state is still preferred but solid state is possible as an unfavoured stable state. At $P = P_M$ solidification occurs. At higher pressures solid state is preferred but liquid state can occur as an unfavoured stable state until P_S beyond which only solid state is stable and thus the only possible state of the system.

These simple backward bifurcations in two separate physical parameter directions can be expressed in a combined way by using bifurcation theory. The relevant normal form (see Table 5.1, form (8) in [14]) for this tricritical bifurcation is $H_0 = x^5 + 2m\lambda x^3 - \lambda^2 x$ and its universal unfolding is

$$H = x^5 + 2m\lambda x^3 - \lambda^2 x + \alpha x + \beta x^3 \qquad (6)$$

where l is the bifurcation parameter, a and b are the unfolding parameters and m the modal parameter. Bifurcations are assumed to be perfect. The equation $H=0$ is equivalent to Eq. (2) if one identifies

$$g = 2m\lambda + \beta \qquad (7)$$

and

$$\varepsilon = -\lambda^2 + \alpha. \qquad (8)$$

For some materials melting curve in the (P, T) plane is expected to have a horizontal tangent at high P. Here the starting point of this tangent is identified as the tricritical point P_{tc}, T_{tc}. Thus g axis in the Landau theory can be identified to be parallel to P axis. In [12] e axis was assumed to be parallel to T axis. Here e is allowed to depend on both T and P. This dependence can be found by taking the critical temperatures to depend quadratically on P since that is the highest power relationship between e and g on the critical lines in the Landau theory. Then all three critical lines can be expressed as

$$T_{tc} - T = a_i(P - P_{tc})^2, i = 1,2,3. \tag{9}$$

When i=1 we have the curve where $\varepsilon = 0$ and $T = T_L(P)$. Denoting by T_{L0} the value of T_L at $P = 0$, one finds a_1 so that

$$T_{tc} - T_L(P) = (T_{tc} - T_{L0})(P/P_{tc} - 1)^2. \tag{10}$$

Moving all the terms to one side allows one to write generally

$$\varepsilon = T_{tc} - T - (T_{tc} - T_{L0})(P/P_{tc} - 1)^2 \tag{11}$$

which is in the form (8) $a = T_{tc} - T$ and

$$\lambda = \sqrt{T_{tc} - T_{L0}}(P/P_{tc} - 1) \tag{12}$$

since $\lambda < o$. Now one can simplify (7) to

$$g = 2m\lambda \tag{13}$$

since $g = 0$ at $P = P_{tc}$. For first order transitions g needs to be negative. So for λ as above, m needs to be positive.

When $i = 2$ we have the melting curve: $T = T_M(p)$. At $P = 0$, $T_M = T_0$ gives a_2. Thus the equation of the melting curve is

$$T_M(P) = T_{tc} - (T_{tc} - T_0)(P/P_{tc} - 1)^2. \tag{14}$$

Inserting this in Eq. (11) one obtains

$$\varepsilon_M = (T_{L0} - T_0)(P/P_{tc} - 1)^2. \tag{15}$$

Combining this with (4) and (13) one can find

$$m = 2\sqrt{(T_{L0} - T_0)/[3(T_{tc} - T_{L0})]}. \tag{16}$$

When $i = 3$ we have the third critical curve marking the end of the hysteresis, the condition for the lowest possible liquid phase temperature T_S. The liquid state must vanish at $T = 0$, so at $P = 0, T_S = 0$ gives $a_3 = T_{tc}/P_{tc}^2$. Thus the equation of the hysteresis curve is

$$T_S(P) = T_{tc} - T_{tc}(P/P_{tc} - 1)^2 \tag{17}$$

and inserting this in (11) one finds

$$\varepsilon_S = T_{L0} \left(P/P_{tc} - 1\right)^2. \tag{18}$$

Using this with (5) one obtains $T_{LO} = 4T_0$ allowing to write the equation (10) as

$$T_L(P) = T_{tc} - (T_{tc} - 4T_0)\left(P/P_{tc} - 1\right)^2. \tag{19}$$

Thus one finds

$$g = 4\sqrt{T_0}\left(P/P_{tc} - 1\right) \tag{20}$$

and

$$\varepsilon = T_{tc} - T - (T_{tc} - 4T_0)\left(P/P_{tc} - 1\right)^2 \tag{21}$$

and

$$x = 0 \text{ or } x = \pm\sqrt{2\sqrt{T_0}\left(1 - P/P_{tc}\right) \pm \sqrt{T - T_{tc} + T_{tc}\left(P/P_{tc} - 1\right)^2}}. \tag{22}$$

These values of x are drawn in Fig. 2.2 using the iron tricritical point obtained from the experimental data as discussed next. In Fig. 2.2a $x(T)$ is shown for $P=329$ GPa (corresponding to the pressure on Earth's inner core boundary [15] which is a solidification front in iron-rich core melt). In Fig. 2.2b the bifurcation structure of $x(P)$ is shown for $T=7500$ K which is greater than T_{LO}, thus exhibiting all three critical transitions. The critical curves (14), (17) and (19) are drawn in Fig. 3.1 with iron melting data and *ab initio* calculations.

3. Experimental evidence for iron

Iron is the dominant element in terrestrial planetary cores. For instance, the Earth has an iron-rich core at depths below about half of the Earth's radius. The outer core is molten but the inner core is close to pure iron which is solidifying out from the outer core melt. Owing to its significant geophysical interest, iron is the most studied high pressure material. In Fig. 3.1, data since 1986 is presented with the *ab initio* calculations for iron melting. For discussion, see [12]. Overall, the data has a large scatter, and all shock wave results as well as the older static measurements have very large error bars. However, selecting (details in [12]) the most reliable static results combined with all supporting high pressure shock results allows one to find an excellent fit to Eq. (14) giving the tricritical point as (793 GPa, 8632 K) with a correlated uncertainty of about $\pm$ (100 GPa, 800 K). The

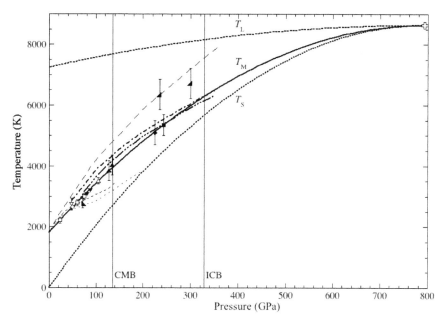

Fig. 3.1. Iron melting experimental data, with *ab initio* calculations without (dash-dotted) and with (dash-triple-dotted) a free energy correction [16] and three theoretical critical lines T_S, T_M and T_L from this work. The most reliable static data (open symbols: squares [17], diamond [18] and triangle [19]) with consistent shock wave data (filled symbols: square [20], diamond [21], triangles [22] and inverted triangles [23]) are used to find the theoretical curve for T_M (solid line) and thus the coordinates of the tricritical point (P_{tc}, T_{tc}) (cross). Other data (see discussion in [12]) are also shown: short dashed [24] and dashed [25] lines and open inverted triangle [26] are for static data, filled symbols are for shock data (lower right corner triangles [27], lower left corner triangle [28]), but the long dashed line [29] represents static data extrapolated to high pressures using shock data. The vertical lines show the pressures at the Earth's core mantle boundary (CMB) and inner core boundary (ICB).

theoretical melting curve (14) is drawn in Fig. 3 and it goes approximately through the middle of the data. In addition, the curves for T_L from Eq. (19) and T_S from Eq. (17) are also drawn showing the limits of the range where unfavoured liquid or solid stable states can occur. All the experimental data are in this range.

The pressure at the Earth's core-mantle boundary (CMB) is about 136 GPa [15]. The iron melting temperature is at that pressure 3945 ± 12 K using Eq. (14). This temperature is very similar to the seismic estimate 3950 ± 200 K [30] for temperature there implying the melt is rather close to pure iron. This agrees very well with the result that experimentally, pure iron compressional wave velocities are very similar to seismic velocities in the Earth's core melt close to the CMB (see Fig. 2 in [31]). However, the mantle solid at the CMB has density of about 56 % of the core melt density there [15]. Thus it is the light matter in the core melt which is solidifying out there, presumably with some iron. This solidification temperature stays very close to pure iron for small concentrations of light matter in the iron-rich melt. Solidification at CMB has been suggested previously [32-34].

The iron melting temperature at the pressure of the Earth's inner core boundary (ICB) is 6290 ± 80 K. This is very close to the *ab initio* result without free energy correction (see Fig. 3.1). However, the inner core is solidifying from the molten iron-rich outer core, which owing to the seismic density estimates [15] there is concluded to have also some light elements. These light impurities are lowering the temperature from the pure iron melting temperature. From this work we can conclude from Eq. (17) that in the real Earth the temperature at the ICB is possible to be as low as T_S, about 5670 K, but not lower. Since the fluid in the outer core has been convecting for billions of years it has been able to adjust its fluid concentration and temperature profiles so that the temperature and density gradients inside the fluid are minimized. Thus the temperature at the ICB is expected to take this limiting value of about 5670 K. This estimate agrees very well with the value of 5700 K as inferred for the temperature at the ICB from *ab initio* calculations on the elasticity properties of the inner core [35] and is consistent with the range 5400 K – 5700 K reported in [36]. The temperature difference $T_M - T_S = 621$ K at ICB also agrees with the estimate 600 K to 700 K in [36] but is more than twice the 300 K used in the rather recent energy budget calculations of the core [37].

4.   Conclusions

The concept of tricritical point seems to be very useful in considering liquid-solid phase transitions as a function of pressure and temperature. Landau theory gives a quadratic, general formula for the solidification/melting curve $T_M(P)$ if it ends at a tricritical point where $dT_M/dP = 0$. The structure of the coefficients is definite, but the tricritical point needs to be estimated from good solidification or melting data. For pure iron, the temperature formula and data agree very well in the whole range 0–250 GPa where we have good experimental data. The tricritical point is estimated to be at (800±100 GPa, 8600±800 K), with the signs of the errors correlated. The prediction for the iron melting temperature at the core-mantle boundary is 3945±12

K, very similar to the seismic estimate for temperature of 3950±200 K at the CMB implying the melt there is rather close to pure iron. The impurities present in the outer core decrease the inner core boundary temperature from 6290±80 K found for pure iron to about 5670 K for the real Earth, in agreement with *ab initio* predictions.

References

1. L. Landau, *Phys. Z. Sowjetunion* **11**, 26 (1937); reprinted in *Collected Papers of L. D. Landau*, ed. D. ter Haar, p. 193 (Pergamon, Oxford 1965).
2. R. B. Griffiths, *Phys. Rev. Lett.* **24**, 715 (1970).
3. I. D. Lawrie and S. Sarbach in *Phase Transitions and Critical Phenomena* **Vol. 9**, eds C. Domb and J. L. Lebowitz, p. 1 (Academic Press, London 1984).
4. C. M. Knobler and R. L. Scott in *Phase Transitions and Critical Phenomena* **Vol. 9**, eds C. Domb and J. L. Lebowitz, p. 163 (Academic Press, London 1984).
5. E. Zeldov, D. Majer, M. Konczykowski, V. B. Geshkenbein, V. M. Vinokur and H. Shtrikman, *Nature* **375**, 373 (1995).
6. J. Thoen, H. Marynissen and W. Van Dael, *Phys. Rev. Lett.* **52**, 204 (1984).
7. Y.-I. Shin, C. H. Schunck, A. Schirotzek and W. Ketterle, *Nature* **451**, 689 (2008).
8. P. S. Peercy, *Phys. Rev. Lett.* **35**, 1581 (1975).
9. R. J. Birgeneau, G. Shirane, M. Blume and W. C. Koehler, *Phys. Rev. Lett.* **33**, 1098 (1974).
10. C. W. Garland and B. B. Weiner, *Phys. Rev.* **B3**, 1634 (1971).
11. Y. P. Feng and M. H. W. Chan, *Phys. Rev. Lett.* **64**, 2148 (1990).
12. A. Aitta, *J. of Stat. Mech.* **2006**, P12015 (2006).
13. A. Aitta, *Phys. Rev.* **A34**, 2086 (1986).
14. M. Golubitsky and D. G. Schaeffer, *Singularities and groups in bifurcation theory* **Vol. I** (Springer-Verlag, New York, 1985).
15. A. M. Dziewonski and D. L. Anderson, *Phys. Earth Planet. Inter.* **25**, 297 (1981).
16. D. Alf, G. D. Price and M. J. Gillan, *Phys. Rev.* **B65**, 165118 (2002).
17. G. Shen, H.-K. Mao, R. J. Hemley, T. S. Duffy and M. L. Rivers, *Geophys. Res. Lett.* **25**, 373 (1998).
18. G. Shen, V. B. Prakapenka, M. L. Rivers and S. R. Sutton, *Phys. Rev. Lett.* **92**, 185701 (2004).
19. Y. Ma, M. Somayazulu, G. Shen, H.-K. Mao, J. Shu and R. J. Hemley, *Phys. Earth Planet. Inter.* **143–144**, 455 (2004).
20. J. M. Brown and R. G. McQueen, *J. Geophys. Res.* **91**, 7485 (1986).
21. J. H. Nguyen and N. C. Holmes, *Nature* **427**, 339 (2004).
22. H. Tan, C. D. Dai, L. Y. Zhang and C. H. Xu, *Appl. Phys. Lett.* **87**, 221905 (2005).
23. Y.-H. Sun, H.-J. Huang, F.-S. Liu, M.-X. Yang and F.-Q. Jing, *Chin. Phys. Lett.* **22**, 2002 (2005).

24. R. Boehler, *Nature* **363**, 534 (1993).
25. S. K. Saxena, G. Shen and P. Lazor, *Science* **264**, 405 (1994).
26. A. P. Jephcoat and S. P. Besedin, *Phil. Trans. R. Soc. Lond. A* **354**, 1333 (1996).
27. C. S. Yoo, N. C. Holmes, M. Ross, D. J. Webb and C. Pike, *Phys. Rev. Lett.* **70**, 3931 (1993).
28. T. J. Ahrens, K. G. Holland and G. Q. Chen, *Geophys. Res. Lett.* **29**, 1150 (2002).
29. Q. Williams, R. Jeanloz, J. Bass, B. Svendsen and T. J. Ahrens, *Science* **236**, 181 (1987).
30. R. D. Van der Hilst, M. V. de Hoop, P. Wang, S.-H. Shim , P. Ma and L. Tenorio, *Science* **315**, 1813 (2007).
31. J. Badro, G. Fiquet, F. Guyot, E. Gregoryanz, F. Occelli, D. Antonangeli and M. d'Astuto, *Earth Planet. Sci. Lett.* **254**, 233 (2007).
32. H. H. Schloessin and J. A. Jacobs, *Can. J. Earth Sci.* **17**, 72 (1980).
33. B. A. Buffett, E. J. Garnero and R. Jeanloz, *Science* **290**, 1338 (2000).
34. B. Buffett, E. Garnero and R. Jeanloz, *Science* **291**, 2091 (2001).
35. G. Steinle-Neumann, L. Stixrude, R. E. Cohen and O. Glseren, *Nature* **413**, 57 (2001).
36. D. Alf, M. J. Gillan and G. D. Price, *Contemp. Phys.* **48**, 63 (2007).
37. J. R. Lister, *Earth Planet. Inter.* **140**, 145 (2003).

ELASTIC WAVES IN RODS OF
RECTANGULAR CROSS SECTION

A. A. BONDARENKO

Department of Natural Sources, Institute of Telecommunications and
Global Information Space of the NAS Ukraine
Kiev, 03186, Ukraine
anastasiya.bondarenko@gmail.com

This article presents an analytical solution for harmonic waves in a rectangular elastic rod of arbitrary aspect ratio. The solution is obtained as a sum of two series, each term of which identically satisfies equations of motion, and has sufficient arbitrariness for fulfilment of any assigned boundary conditions on the rod surface. Because of interdependency of the series coefficients, dispersion relation is obtained in the form of an infinite determinant. Correct reduction of the determinant permits to establish the edge effect on the wave dispersion distinguishing a rectangular rod from an infinite plate and a circular cylinder. Calculated results excellently agree with Morse's experimental data (1948).

Keywords: Rectangular rod; longitudinal waves; dispersion.

1. Introduction

Ultrasonic volume and surface waves in elastic rods of various cross-section are extensively used to solve a number of important practical problems. Originally their applications included delay lines, frequency filters, and flaw detection. Recently tremendous potential has been uncovered by application of these waves for nondestructive material evaluation [1], theoretical modeling of transport phenomena in nanometre-scale wires [2], and mixing of viscous solutions in microchannels at low Reynolds number [3].

Exact solutions of the equations of motion can be derived, however, only for isotropic infinite plates and circular cylinders due to two-dimensionality of the wave field in such simple geometries. For rods of other cross-section, the problem becomes substantially three-dimensional. The presence of breaks at the elastic rod surface causes considerable complication of the wave field structure owing to additional reflections of compressional and shear waves forming normal modes. To analyze the surface effect on the wave propagation process, it is advisable to study dispersion characteristics for a rod of rectangular cross-section, which is also the base element for rods of arbitrary cross-section [2].

Two sets of the exact solutions for a rectangular rod were discovered by Lamé (1852) and Mindlin & Fox (1960) for certain discrete frequencies and cross-sectional aspect ratio. Numerous attempts were made to obtain general solution of the problem for rods with arbitrary aspect ratio. They resulted in development of different approximate theories based on simplified hypothesis for a rod stress-strain state (for relevant literature, see, for example, [4, 5]). Most of them are valid only in a low frequency range and for rods of either large aspect ratio or square cross-section. Therefore, these theories fail to interpret extensive experimental data on the elastic wave dispersion in a wide range of rectangular cross-sectional rods presented by Giebe & Blechshmidt [6] and Morse [7, 8]. To the author's knowledge, so far there are no reliable solutions to model this data.

In this paper, a complete analytical solution to the problem is presented for rectangular rods with arbitrary aspect ratio. The solution consists of two infinite sets of partial solutions for equations of motion, which are able to satisfy any assigned boundary conditions at the rod surface. Dispersion equation is established in terms of an infinite determinant. Reduction of the determinant is based on the knowledge of asymptotic behavior of unknown coefficients in the system. Excellent agreement of theoretical predictions on wave dispersion for two lowest longitudinal modes with experimental data described by Morse [7] proves a high accuracy of the results obtained. Calculations performed in a high frequency range reveal the edge effect on the wave field that results in decrease of the limiting values for phase velocities and redistribution of wave motions towards the rod edges. Characteristics established substantially distinguish a rectangular rod from an infinite plate and a cylinder, and are of great importance for many practical applications.

2. Formulation of the problem

Small motions of an isotropic elastic solid characterized by density ρ and Lamé constants λ, μ are described by vector equation of motion

$$\mu \nabla^2 \vec{U} + (\lambda + \mu)\nabla(\nabla \cdot \vec{U}) = \rho \partial^2 \vec{U}/\partial t^2, \tag{1}$$

where t is time, $\vec{U}(x, y, z, t)$ is displacement vector.

The vector equation (1) has a disadvantageous feature in that it couples the three displacement components. It is far more convenient to express the components of the displacement vector in terms of the scalar Φ and vector

$\vec{\Psi}$ potentials as follows [9]

$$\vec{U} = \nabla\Phi + (\nabla \times \vec{\Psi}), \quad \nabla \cdot \vec{\Psi} = 0. \tag{2}$$

The potential introduced satisfy uncoupled Helmholtz wave equations. To determine a set of normal waves appropriate to a rectangular rod $-a \leq x \leq a$, $-b \leq y \leq b$ infinite along z axis, potential functions should also satisfy the boundary conditions

$$\sigma_x = \tau_{xy} = \tau_{xz} = 0 \quad \text{at} \quad x = \pm a,$$
$$\sigma_y = \tau_{yx} = \tau_{yz} = 0 \quad \text{at} \quad y = \pm b. \tag{3}$$

Substitution of Eq.(2) into Hooke's law provides expressions for normal σ_i and shear τ_{ij} stresses in terms of the potential functions.

In case of progressive harmonic waves, the potential functions are assumed to be given in the form $\{\Phi, \vec{\Psi}\}(x, y, z, t) = \{\phi, \vec{\psi}\}(x, y) \exp i(\gamma z - \omega t)$, where ω is the frequency, γ is the propagation constant ($\gamma = 2\pi/\lambda$, λ is the wavelength). In what follows, the factor $\exp i(\gamma z - \omega t)$ will be omitted.

3. Method of solution

For longitudinal waves symmetric relative to the middle planes of the rod, consider following expressions for the potentials as solutions of Eqs. (1), (2)

$$\phi = A \cosh p_1 y \cos \alpha x, \quad \psi_x = B \sinh p_2 y \cos \alpha x,$$
$$\psi_y = C \cosh p_2 y \sin \alpha x, \quad \psi_z = D \sinh p_2 y \sin \alpha x, \tag{4}$$

with the notation

$$p_1^2 = \alpha^2 + \gamma^2 - \Omega_1^2, \quad p_2^2 = \alpha^2 + \gamma^2 - \Omega_2^2, \quad \Omega_1 = \frac{\omega}{c_1}, \quad \Omega_2 = \frac{\omega}{c_2};$$

c_1 and c_2 are velocities of compressional and shear waves, respectively. Here A, B, C, D are unknown constants determined from boundary conditions (3) and related as $Di\gamma + Cp_2 - \alpha B = 0$ according to the second equation in (2). When $p_1^2 < 0$ or $p_2^2 < 0$, hyperbolic functions in (4) are replaced by trigonometric ones.

The values of α are chosen in such a way that trigonometric functions $\{\sin \alpha x, \cos \alpha x\}$ constitute complete and orthogonal sets of functions at $-a \leq x \leq a$, for example,

$$\alpha_n = \frac{n\pi}{a}, \quad n = 0, 1, 2, \dots.$$

Therefore, from function representations (4) we obtain an infinite set of partial solution of Eqs. (1) that enables satisfying any assigned boundary conditions on the faces $y = \pm b$ by means of matching the values of A_n, B_n, and C_n. Another infinite set of partial solutions for fulfilment the boundary conditions at $x = \pm a$ is obtained from the condition of completeness and orthogonality for sets of trigonometric functions $\{\sin \beta y, \cos \beta y\}$ at $-b \leq y \leq b$:

$$\phi = \sum_{k=0}^{\infty} E_k \cosh q_1 x \cos \beta_k y, \quad \psi_x = \sum_{k=0}^{\infty} F_k \cosh q_2 x \sin \beta_k y,$$

$$\psi_y = \sum_{k=0}^{\infty} G_k \sinh q_2 x \cos \beta_k y, \quad \psi_z = \sum_{k=0}^{\infty} H_k \sinh q_2 x \sin \beta_k y, \quad (5)$$

where

$$\beta = \frac{k\pi}{b}, \quad q_1^2 = \beta_k^2 + \gamma^2 - \Omega_1^2, \quad q_2^2 = \beta_k^2 + \gamma^2 - \Omega_2^2.$$

A sum of the two infinite sets provides general solution to the problem for longitudinal waves in a rod of arbitrary rectangular cross section. Corresponding expressions for displacements and stresses are rather cumbersome and not cited here to save space.

Fulfilment of conditions (3) for shear stresses leads to the relations

$$i\gamma A_n p_1 \sinh p_1 b - B_n \frac{p_2^2 + \alpha_n^2 + \gamma^2}{2} \sinh p_2 b = 0, \quad C_n = 0;$$

$$i\gamma E_k q_1 \sinh q_1 a + G_k \frac{q_2^2 + \beta_k^2 + \gamma^2}{2} \sinh q_2 a = 0, \quad F_k = 0.$$

It is easily seen that there remain two sets of arbitrary constants, for example, B_n and G_k. While satisfying boundary conditions (3) for normal stresses, we obtain two functional equations for determining these constants. Taking into account the Fourier expansions for hyperbolic functions into the trigonometric series, after some computations those equations are converted into the infinite system of linear algebraic equations

$$Y_k a \Delta_k(q) + \varepsilon_k \sum_{n=0}^{\infty} X_n b_n \left[\frac{2\alpha_n^2 \beta_k^2}{\alpha_n^2 + q_1^2} - \frac{2\alpha_n^2 \beta_k^2}{\alpha_n^2 + q_2^2} - \frac{\Omega_0^2 \left(2\gamma^2 - \Omega_2^2\right)}{\alpha_n^2 + q_1^2} \right] = 0,$$

$$X_n b \Delta_n(p) + \varepsilon_n \sum_{k=0}^{\infty} Y_k c_k \left[\frac{2\alpha_n^2 \beta_k^2}{\beta_k^2 + p_1^2} - \frac{2\alpha_n^2 \beta_k^2}{\beta_k^2 + p_2^2} - \frac{\Omega_0^2 \left(2\gamma^2 - \Omega_2^2\right)}{\beta_k^2 + p_1^2} \right] = 0,$$

$$k, n = 0, 1, 2, \dots . \quad (6)$$

Here new coefficients X_n, Y_k are related to B_n, G_k as follows

$$X_n = (-1)^{n-1} \frac{B_n}{i\gamma b b_n} \sinh p_2 b, \quad Y_k = (-1)^{k-1} \frac{F_k}{i\gamma a c_k} \sinh q_2 a,$$

and the notations

$$\Delta_k(q) = c_k \left\{ q_2 \left(\gamma^2 + \beta_k^2\right) \coth q_2 a - \frac{\left(\gamma^2 + \beta_k^2 + q_2^2\right)^2}{4q_1} \coth q_1 a \right\},$$

$$\Delta_n(p) = b_n \left\{ p_2 \left(\gamma^2 + \alpha_n^2\right) \coth p_2 b - \frac{\left(\gamma^2 + \alpha_n^2 + p_2^2\right)^2}{4p_1} \coth p_1 b \right\};$$

$$\varepsilon_i = \begin{cases} \frac{1}{2}, & i = 0, \\ 1, & i > 0; \end{cases} \quad c_k = \begin{cases} 1, & k = 0, \\ \frac{1}{\beta_k^2}, & k > 0; \end{cases} \quad b_n = \begin{cases} 1, & n = 0, \\ \frac{1}{\alpha_n^2}, & n > 0; \end{cases}$$

$\Omega_0^2 = \nu \Omega_1^2/(1 - 2\nu)$, $\nu = \lambda/2(\lambda + \mu)$ are introduced.

The only non-trivial solutions for X_n, Y_k are those, for which the determinant of system (6) is equal to zero. The equation formed by expansion of the determinant is a dispersion equation, which for a given value of a/b relates frequency ω to propagation constant γ with Poisson's ratio ν as a parameter. For any positive real values of ω, roots of the dispersion equation are real, imaginary and complex values of γ corresponding to normal modes in the rectangular rod.

For proper reduction of the infinite system to a finite one, an important role plays the law of asymptotic behavior of the unknowns for large values of indices [10]

$$\lim_{n \to \infty} X_n = \lim_{k \to \infty} Y_k = A, \tag{7}$$

where A is an unknown constant; its value depends on frequency ω. On a basis of law (7), the improved reduction method, according to which a finite system consists of $N + K + 3$ equations for the same number of unknown coefficients X_n $(n = 0, 1, \ldots, N)$, Y_k $(k = 0, 1, \ldots, K)$, and A, was suggested. An equation for determining A can be easily obtained after some transformations of the infinite system. Detailed description of this procedure and its mathematical background are given in [10]. It is important to note that the application of the improved reduction method permits to considerably increase the accuracy in finding dispersion characteristics of normal waves compared to a simple reduction method, when the value of A is supposed to be zero.

4. Results and discussion

Since the dispersion equation is a transcendental one, it has an infinite number of roots γ for each value of ω. Roots constitute dispersion curves in the (ω, γ) space corresponding to propagating (real γ) and non-propagating (imaginary and complex γ) normal modes in the rod. Propagating modes are characterized by phase $c_p/c_2 = \omega/\gamma$ and group $c_g/c_2 = d\omega/d\gamma$ velocities.

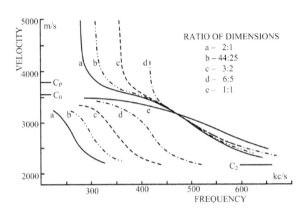

Fig. 4.1. Two lowest longitudinal modes in rectangular brass rods of various aspect ratio observed experimentally by Morse (reproduced from Fig. 5 in paper [7]).

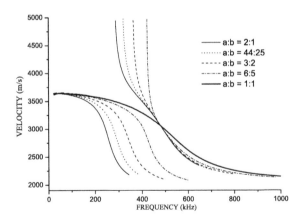

Fig. 4.2. Two lowest longitudinal modes in rectangular rods of various aspect ratio calculated theoretically by means of the theory developed, $\nu = 0.35$.

Morse observed experimentally phase velocities of two lowest longitudinal waves on seven brass rods of various aspect ratio [7]. All of the rods have one common lateral dimension, $2b = 0.318$ cm. Fig. 4.1 summarizes Morse's results as the larger side of the rod, $2a$, is varied. Theoretical predictions are presented in Fig. 4.2 on the same scale as experimental data. Calculations were implemented for Poisson's ratio $\nu = 0.35$, and shear wave velocity $c_2 = 2160$ m/s as suggested in [7].

The first longitudinal mode extends the zero frequency and is almost non-dispersion at low frequencies, $0 < f < 100$ kHz. Regardless of aspect ratio, its phase velocity equals bar velocity $c_0 = E/\rho = 3651$ m/s. This behavior was not observed by Morse, because the lower end of the frequency range was limited by the oscillator employed [7]. As frequency is increased, the phase velocity monotonically decreases and approaches, according to Morse's prediction, Rayleigh's surface wave velocity $c_R = 2019$ m/s. Observations in a high frequency range were limited by difficulty in distinguishing the nodal lines when the wavelength was less than 3 mm.

The second mode has a non-zero cut-off frequency, at which the motion is independent of the z coordinate. Morse's measurements show that at the cut-off frequency the half-wavelength is approximately equal to the larger side $2a$. This conclusion is also valid for the calculated values of cut-off frequencies that exceed less than 0.5% the values of frequencies, at which the half-wavelength equals the side $2a$. With increasing frequency, the phase velocity of this mode drops from infinity to a certain finite value in monotone fashion. All the curves describing the second mode have a common point at frequency $f_L = 484$ kHz, at which the half-wave-length equals the smaller side $2b$. Theoretically this frequency corresponds to the exact Lamé solution, and its location is independent of Poisson's ratio ν and aspect ratio a/b.

Comparison of calculated results with experimental data shows an excellent agreement in the whole frequency range that confirms the efficiency of the theory developed. The solution obtained permits also to evaluate the high frequency limits for phase velocities of the two modes under consideration. The knowledge of those limits is of importance both from the practical point of view and for theoretical interpretation of the wave propagation process in rectangular rods.

Calculations show that phase velocity of the lowest longitudinal mode regardless of the aspect ratio asymptotically approaches the velocity of edge angular mode $c_E = 1957$ m/s for a right wedge [11] as frequency is increased. At these frequencies wave motions are concentrated near the edges of the

Table 4.1.   Forms of rod cross section for the first longitudinal mode ($a/b = 1$, $\nu = 0.35$).

$\lambda = 12.23$ mm	$\lambda = 2.48$ mm	$\lambda = 1.18$ mm	$\lambda = 0.63$ mm

Table 4.2.   Phase-velocities for two lowest longitudinal waves in a rectangular rod ($a/b = 2$, $\nu = 0.35$).

	Simple reduction		Improved reduction	
λ, mm	$c_p^{(1)}$, m/s	$c_p^{(2)}$, m/s	$c_p^{(1)}$, m/s	$c_p^{(2)}$, m/s
1.272	1960	2032	1957	2032
0.636	1962	2022	1958	2022
0.424	1961	2021	1958	2020
0.318	1962	2020	1957	2019

rod (see Table 4.1). This fact essentially distinguishes a rectangular rod from an infinite plate or a cylinder, for which a high frequency limiting value for phase velocity of the first mode is c_R, and wave motions are distributed along the surfaces like a Rayleigh wave. Similarly, with increasing frequency, phase velocity of the second longitudinal mode approaches the value of c_R, rather than c_2 as in the case of the plate or cylinder. This behavior can be explained by the presence of breaks on the rod surface, since wave motions are located near the edges when the wavelength becomes much smaller than any of the rod dimension.

Table 4.2 presents a comparison between the two reduction approaches to the solution of system (6). Calculations in both cases were performed for the brass rod with aspect ratio $a/b = 2$ when $N = 20$, $K = 10$ in corresponding finite systems. It is easily seen that the simple reduction method gives slightly conservative values for phase velocity compared to the improved reduction. This small difference could hardly be observed experimentally. Nevertheless, the improved reduction method permits to establish more accurate limiting values and to catch distinctive features in the wave field distribution shown in Table 1.

5. Conclusion

An analytical solution for normal waves in rectangular rods of arbitrary cross section is presented. Dispersion equation obtained in terms of an infinite determinant is solved taking into account asymptotic behavior of unknowns in the system. The improvement of the reduction method for the infinite system results in establishing more accurate high-frequency limiting values for phase velocities of two lowest longitudinal modes. Whereas these values are slightly smaller than those for an infinite plate or a circular cylinder, wave motion distribution differs significantly. These results could be of great importance for many practical applications, especially, for spacing exciting and receiving transducers by nondestructive material evaluation.

Acknowledgement

The author would like to acknowledge the guidance of Prof. V. V. Meleshko (Kiev National Taras Shevchenko University, Ukraine) who supervised the present work.

References

1. T. Hayashi, W.-J. Song and J. L. Rose, *Ultrasonics*, **41**, 175 (2003).
2. N. Nishiguchi, Y. Ando and M. N. Wybourne, *J. Phys.: Condens. Matter* **9**, 5751 (1997).
3. A. D. Stroock, S. K. W. Dertinger, A. Ajdari, I. Mezi, H. A. Stone and G. M. Whitesides.*Science* **295**, 647 (2002).
4. T. R. Meeker and A. H. Meitzler, *Guided wave propagation in elongated cylinders and plates*, in *Physical Acoustics. Principles and Methods*, ed. W. P. Mason (New York: Academic, 1964), pp. 111–167.
5. B. A. Auld, *Acoustic Fields ind Waves in Solids* (Krieger, Malabar, FL, 1990).
6. E. Giebe and E. Blechschmidt, *Ann. Physik* **18**, 457 (1933).
7. R. W. Morse, *J. Acoust. Soc. Amer.* **20**, 833 (1948).
8. R. W. Morse, *J. Acoust. Soc. Amer.* **22**, 219 (1950).
9. J. D. Achenbach, *Wave Propagation in Elastic Solids* (North Holland, Amsterdam, 1973).
10. V. T. Grinchenko and V. V. Meleshko, *Harmonic Vibrations and Waves in Elastic Bodies* (Naukova Dumka, Kiev, 1981), in Russian.
11. V. T. Grinchenko and V. V. Meleshko, *Acoust. Phys.* **27**, 206 (1981) in Russian.

NATURAL EXTENSIONS FOR THE GOLDEN MEAN

K. DAJANI

Department of Mathematics, Utrecht University
Utrecht, Postbus 80.000, 3508 TA, the Netherlands
k.dajani1@uu.nl
www.math.uu.nl/people/dajani

C. KALLE

Department of Mathematics, Utrecht University
Utrecht, Postbus 80.000, 3508 TA, the Netherlands
c.c.c.j.kalle@uu.nl
www.math.uu.nl/people/kalle

We give two versions of the natural extension of a specific greedy β-transformation with arbitrary digits. We use the natural extension to obtain an explicit expression for the invariant measure, equivalent to the Lebesgue measure, of this β-transformation.

Keywords: Greedy expansion; natural extension; equivalent invariant measure.

1. Introduction

Real numbers can be represented in many different ways. Famous examples are the integer base expansions and the continued fraction expansions. Another well-studied example is given by the β-expansion, which is an expression of the form $x = \sum_{n=1}^{\infty} b_n/\beta^n$, where $\beta > 1$ is a real number and the digits b_n are integers between 0 and the largest integer smaller than β. In 1957 Rényi [1] introduced the transformation $T_\beta\, x = \beta x \pmod 1$, that generates such expansions by iteration. The introduction of this transformation made it possible to use ergodic theory to study β-expansions.

In this article we consider a specific piecewise linear transformation with constant slope, which falls into the class of greedy β-transformations with arbitrary digits as defined in (D.,K. 07) [2]. The transformation T_β, as given by Rényi, is also contained in this class. We can use this more general class of transformations to obtain β-expansions with digits in arbitrary sets of real numbers, satisfying a mild condition. A recursive algorithm to produce

β-expansions with general digit sets was first given by Pedicini [3] and some properties of the greedy β-transformations with arbitrary digits are given in (D., K. 09) [4]. In the next section, we consider the case $\beta = (1 + \sqrt{5})/2$, the golden mean, and show how one can obtain β-expansions with digits 0, 2 and 3. These methods can be generalized to arbitrary $\beta > 1$ [2].

As said before, the advantage of having a transformation to generate expansions, is that we can use ergodic theory as a tool. For that, we first need an invariant measure for the underlying transformation. In general, a β-transformation is not invertible. A way to obtain an invariant measure is by constructing an invertible dynamical system that contains the dynamics of T and is minimal from a measure theoretical point of view. Such an invertible system is called a *natural extension*. It is a basic object in ergodic theory and can be used to obtain many properties of the transformation and the expansions it generates. An invariant measure of T can be found through an invariant measure of a natural extension. A precise definition of a natural extension is given by the conditions (i)-(iv) from Section 3.

There is a canonical way to construct natural extensions [5] [6]. In general, there are many ways to construct such an invertible system, but it is proven by Rohlin [5] that any two such systems are isomorphic. So, we can speak of the natural extension of a dynamical system up to isomorphism. If we want to use the natural extension to prove properties of a transformation, then the success of the whole process depends heavily on the version of the natural extension that we choose to work with.

In this paper, we consider a special transformation T, generating β-expansions with β the golden mean and digits 0, 2 and 3. We give two versions of the natural extension of T. The first version that we define, will prove to be invariant with respect to the Lebesgue measure. This allows us to give an invariant measure for T, equivalent to the Lebesgue measure, by simply projecting the measure of the natural extension. The second version of the natural extension is a planar one and thus easier to visualize. Versions of the natural extension for similar transformations are given by Dajani et al. [7] and by Brown and Yin [8]. The first version we give is a generalization of the natural extension defined in (Dajani et al., 1996) [7]. In the next section of this paper we define the transformation T and give some of its properties. In the third section we give a first version of the natural extension and the density of the invariant measure. The fourth section is used to give the planar version of the natural extension and to establish an isomorphism between the two versions.

2. Expansions and fundamental intervals

Let β be the golden mean, i.e. the positive solution of the equation $x^2 - x - 1 = 0$. Consider the partition $\Delta = \{\Delta(0), \Delta(2), \Delta(3)\}$ of the interval $[0, 2)$, given by $\Delta(0) = [0, 2/\beta)$, $\Delta(2) = [2/\beta, 3/\beta)$ and $\Delta(3) = [3/\beta, 2)$. The transformation $T : [0, 2) \to [0, 2)$ is defined by

$$Tx = \beta x - j, \quad \text{if } x \in \Delta(j). \tag{1}$$

To each $x \in [0, 2)$ we assign a digit sequence $\{d_n(x)\}_{n \geq 1}$ as follows. Let $d_1(x)$ be

$$d_1(x) = \begin{cases} 0, & \text{if } x \in [0, 2/\beta), \\ 2, & \text{if } x \in [2/\beta, 3/\beta), \\ 3, & \text{if } x \in [3/\beta, 2), \end{cases}$$

and for $n \geq 2$, set $d_n(x) = d_1(T^{n-1}x)$. We can write $Tx = \beta x - d_1(x)$ and inverting this, we get $x = d_1(x)/\beta + Tx/\beta$. By iteration we get for all $n \geq 1$,

$$x = \frac{d_1(x)}{\beta} + \frac{d_2(x)}{\beta^2} + \ldots + \frac{d_n(x)}{\beta^n} + \frac{T^n x}{\beta^n}.$$

Since $T^n x \in [0, 2)$ for all $n \geq 1$, we get that $x = \sum_{n=1}^{\infty} d_n(x)/\beta^n$. So, we can write each x as a β-expansion with digits in $\{0, 2, 3\}$. For ease of notation we sometimes identify x with the infinite sequence $d_1(x)d_2(x)\ldots$. The transformation T generates expansions of all points in the interval $[0, 2)$. Two expansions that will play an important role in what follows are the expansions of the points 1 and β^{-3}. Notice that $\beta^{-3} = 2\beta - 3$ would be the image of 2 under T if T were defined on the closed interval $[0, 2]$. We have

$$1 = \sum_{n=1}^{\infty} \frac{d_n^{(2)}}{\beta^n} = \frac{2}{\beta^2} + \frac{2}{\beta^5} + \frac{2}{\beta^8} + \frac{2}{\beta^{11}} + \ldots = 02\overline{002}, \tag{2}$$

$$\frac{1}{\beta^3} = \sum_{n=1}^{\infty} \frac{d_n^{(3)}}{\beta^n} = \frac{2}{\beta^4} + \frac{2}{\beta^7} + \frac{2}{\beta^{10}} + \ldots = 000\overline{02}, \tag{3}$$

where the bars on the right hand sides indicate repeating blocks of digits. With the *orbit of a point x under T* we mean the set $\{T^n x : n \geq 0\}$. In Figure 2.1, you can see the graph of T and the orbits of the points 1 and β^{-3}.

Using T and Δ, we can define a sequence of partitions $\{\Delta^{(n)}\}_{n \geq 1}$ of $[0, 2)$ by setting $\Delta^{(n)} = \bigvee_{i=0}^{n-1} T^{-i}\Delta$. We call the elements of $\Delta^{(n)}$ *fundamental intervals of rank n*. Since they have the form $\Delta(b_1) \cap T^{-1}\Delta(b_2) \cap \ldots \cap T^{-(n-1)}\Delta(b_n)$ for some $b_1, b_2, \ldots, b_n \in \{0, 2, 3\}$, we will denote them

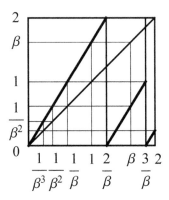

Fig. 2.1.   The transformation T and the orbits of 1 and $1/\beta^3$.

by $\Delta(b_1 \ldots b_n)$. Notice that a fundamental interval of rank n specifies the first n digits of the expansion of the elements it contains. Let λ denote the one-dimensional Lebesgue measure. We will call $\Delta(b_1 \ldots b_n) \in \Delta^{(n)}$ *full* if $\lambda(T^n \Delta(b_1 \ldots b_n)) = 2$ and *non-full* otherwise. $\Delta(0)$ is full and $\Delta(2)$ and $\Delta(3)$ are non-full. By looking at Figure 2.1, we get that after two steps, $\Delta(2)$ splits into a full and a non-full piece, i.e. $\Delta(200)$ is full and $\Delta(202)$ is non-full. The same holds for $\Delta(3)$ after five steps, i.e. $\Delta(300000)$ is full and $\Delta(300002)$ is non-full. After that, each remaining non-full fundamental interval splits into a full and a non-full piece after each three steps: $\Delta(300002000)$ is full, $\Delta(300002002)$ is non-full, $\Delta(202000)$ is full, $\Delta(202002)$ is non-full, etc. For full fundamental intervals, we have the following obvious lemma.

Lemma 2.1. *Let $\Delta(a_1 \ldots a_p)$ and $\Delta(b_1 \ldots b_q)$ be two full fundamental intervals of rank p and q respectively. Then the set $\Delta(a_1 \ldots a_p b_1 \ldots b_q)$ is a full fundamental interval of rank $p + q$.*

A full fundamental interval of rank n has Lebesgue measure $2/\beta^n$. A non-full fundamental interval of the same rank has measure smaller than $2/\beta^n$. From the next lemma, it follows that the full fundamental intervals generate the Borel σ-algebra on $[0, 2)$.

Lemma 2.2. *For each $n \geq 1$, let D_n be the union of those full fundamental intervals of rank n that are not subsets of any full fundamental interval of lower rank. Then $\sum_{n=1}^{\infty} \lambda(D_n) = 2$.*

Proof. Notice that $D_1 = \Delta(0)$, $D_3 = \Delta(200)$ and for $k \geq 1$, D_{3k+1} is the union of two intervals. For all the other values of n, $D_n = \emptyset$. So

$$\sum_{n=1}^{\infty} \lambda(D_n) = \frac{2}{\beta} + \frac{2}{\beta^3} + \sum_{k=1}^{\infty} \frac{4}{\beta^{3(k+1)}} = \frac{2}{\beta} + \frac{2}{\beta^3} + \frac{4}{\beta^3}\left[\frac{1}{1 - 1/\beta^3} - 1\right] = 2.$$

$\square$

Remark 2.1. The fact that $\Delta(0)$ is a full fundamental interval of rank 1 allows us to construct full fundamental intervals of arbitrary small Lebesgue measure. This together with the previous lemma guarantees that we can write each interval in $[0, 2)$ as a countable union of these full intervals. Thus, the full fundamental intervals generate the Borel σ-algebra on $[0, 2)$.

3. Two rows of rectangles

To find an expression for the T-invariant measure, equivalent to Lebesgue, we define two versions of the natural extension of the dynamical system $([0, 2), \mathcal{B}([0, 2)), \mu, T)$. Here T is the transformation as defined in (1). For the definition of the first version, we use a subcollection of the collection of fundamental intervals. For $n \geq 1$, let B_n denote the collection of all non-full fundamental intervals of rank n that are not a subset of any full fundamental interval of lower rank. The elements of B_n can be explicitly given as follows. $B_1 = \{\Delta(2), \Delta(3)\}$, $B_2 = \{\Delta(20), \Delta(30)\}$ and for $k \geq 1$,

$$B_{3k} = \{\Delta(202\underbrace{002\ldots002}_{k-1 \text{ times}}), \Delta(300\underbrace{002\ldots002}_{k-1 \text{ times}})\},$$

$$B_{3k+1} = \{\Delta(202\underbrace{002\ldots002}_{k-1 \text{ times}}0), \Delta(300\underbrace{002\ldots002}_{k-1 \text{ times}}0)\},$$

$$B_{3k+2} = \{\Delta(202\underbrace{002\ldots002}_{k-1 \text{ times}}00), \Delta(300\underbrace{002\ldots002}_{k-1 \text{ times}}00)\}.$$

For each element $\Delta(b_1 \ldots b_n)$ of B_n, $T^n\Delta(b_1 \ldots b_n)$ is one of the intervals $[0, 1/\beta^3)$, $[0, 1/\beta^2)$, $[0, 1/\beta)$, $[0, 1)$ or $[0, \beta)$. The domain of the natural extension will consist of two sequences of sets $\{R_{(2,n)}\}_{n \geq 1}$ and $\{R_{(3,n)}\}_{n \geq 1}$, that represent the images of the elements of B_n under T^n. We will order them in two rows by assigning two extra parameters to each rectangle. Let $R_0 = [0, 2) \times [0, 2) \times \{0\} \times \{0\}$ and for each $n \geq 1$, $j \in \{2, 3\}$ define the sets

$$R_{(j,n)} = T^n\Delta(jd_1^{(j)} \ldots d_{n-1}^{(j)}) \times \Delta(\underbrace{0\ldots0}_{n \text{ times}}) \times \{j\} \times \{n\},$$

where the digits $d_n^{(j)}$ are the digits from the expansions of 1 and $\frac{1}{\beta^3}$ as given in (2) and (3). Then $R = R_0 \cup \bigcup_{n=1}^{\infty}(R_{(2,n)} \cup R_{(3,n)})$. Let $\mathcal{B}_0$ denote the

Borel σ-algebra on R_0 and on each of the rectangles $R_{(j,n)}$, let $\mathcal{B}_{(j,n)}$ denote the Borel σ-algebra defined on it. We can define a σ-algebra on R as the disjoint union of all these σ-algebras, $\mathcal{B} = \coprod_{j,n} \mathcal{B}_{(j,n)} \amalg \mathcal{B}_0$. Let $\bar{\lambda}$ be the measure on $(R, \mathcal{B})$, given by the Lebesgue measure on each rectangle. Then $\bar{\lambda}(R) = 32 - 14\beta$. If we set $\nu = \frac{1}{32-14\beta}\bar{\lambda}$, then $(R, \mathcal{B}, \nu)$ will be a probability space. The transformation $\mathcal{T}$ is defined piecewise on the sets of R. On R_0, let

$$\mathcal{T}(x,y,0,0) = \begin{cases} (Tx, \frac{y}{\beta}, 0, 0) \in R_0, & \text{if } x \in \Delta(0), \\ (Tx, \frac{y}{\beta}, j, 1) \in R_{(j,1)}, & \text{if } x \in \Delta(j),\ j \in \{2,3\} \end{cases}$$

and for $(x, y, j, n) \in R_{(j,n)}$, let

$$\mathcal{T}(x,y,j,n) = \begin{cases} (Tx, y^{(j)}, 0, 0) \in R_0, & \text{if } \Delta(jd_1^{(j)} \ldots d_{n-1}^{(j)} 0) \text{ is full} \\ & \text{and } x \in \Delta(0), \\ (Tx, \frac{y}{\beta}, j, n+1) \in R_{(j,n+1)}, & \text{if } \Delta(jd_1^{(j)} \ldots d_{n-1}^{(j)} 0) \\ & \text{is non-full or } x \notin \Delta(0), \end{cases}$$

where

$$y^{(j)} = \frac{j}{\beta} + \frac{d_1^{(j)}}{\beta^2} + \frac{d_2^{(j)}}{\beta^3} + \ldots + \frac{d_{n-1}^{(j)}}{\beta^n} + \frac{y}{\beta}.$$

Figure 3.1 shows the space R.

Remark 3.1. Notice that for $k \geq 1$, $\mathcal{T}$ maps all rectangles $R_{(2,n)}$ for which $n \neq 3k - 1$ and all rectangles $R_{(3,n)}$ for which $n \neq 3k + 2$ bijectively onto $R_{(2,n+1)}$ and $R_{(3,n+1)}$ respectively. The rectangles $R_{(2,3k-1)}$ and $R_{(3,3k+2)}$ are partly mapped onto $R_{(2,3k)}$ and $R_{(3,3k+3)}$ and partly into R_0. From Lemma 2.2 it follows that $\mathcal{T}$ is bijective.

Let $\pi : R \to [0, 2)$ be the projection onto the first coordinate. To show that $(R, \mathcal{B}, \nu, \mathcal{T})$ is a version of the natural extension with π as a factor map, we need to prove all of the following.

(i) π is surjective, measurable and measure preserving from R to $[0, 2)$.
(ii) For all $x \in R$, we have $(T \circ \pi)(x) = (\pi \circ \mathcal{T})(x)$.
(iii) $\mathcal{T} : R \to R$ is an invertible transformation.
(iv) $\mathcal{B} = \bigvee_{n=0}^{\infty} \mathcal{T}^n \pi^{-1}(\mathcal{B}([0,2)))$, where $\bigvee_{n=0}^{\infty} \mathcal{T}^n \pi^{-1}(\mathcal{B}([0,2)))$ is the smallest σ-algebra containing the σ-algebras $\mathcal{T}^n \pi^{-1}(\mathcal{B}([0,2)))$ for all $n \geq 1$.

The only thing that remains to prove is (iv). For this, we need to have a closer look at the structure of the fundamental intervals and we will introduce some more notation.

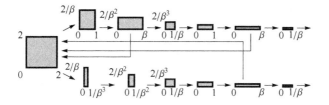

Fig. 3.1.   The space R consists of all these rectangles.

For a fundamental interval, $\Delta(b_1 \ldots b_q)$, we can make a subdivision of the block of digits $b_1 \ldots b_q$ into several subblocks, each of which corresponds to a full fundamental interval, except for possibly the last subblock. This last subblock will form a full fundamental interval if $\Delta(b_1 \ldots b_q)$ is full and it will form a non-full fundamental interval otherwise. To make this subdivision more precise, we need the notion of return time. For points $(x, y) \in R_0$ define the *first return time to R_0* by

$$r_1(x, y) = \inf\{n \geq 1 : T^n(x, y, 0, 0) \in R_0\}$$

and for $k \geq 1$, let the *k-th return time to R_0* be given recursively by

$$r_k(x, y) = \inf\{n > r_{k-1}(x, y) : T^n(x, y, 0, 0) \in R_0\}.$$

Notice that this notion depends only on x, i.e. for all $y, y' \in R_0$ and all $k \geq 1$, $r_k(x, y) = r_k(x, y')$. So we can write $r_k(x)$ instead of $r_k(x, y)$. If $\Delta(b_1 \ldots b_q) \in \Delta^{(q)}$, then for all $n \leq q$, T^n maps the whole set $\Delta(b_1 \ldots b_q) \times [0, 2) \times \{0\} \times \{0\}$ to the same rectangle in R. So the first several return times to R_0, $r_1, \ldots, r_\kappa$, are equal for all $x \in \Delta(b_1 \ldots b_q)$. Then there is a $\kappa \geq 1$ and there are numbers r_i, $1 \leq i \leq \kappa$ such that $r_i = r_i(x)$ for all $x \in \Delta(b_1 \ldots b_q)$. If $\Delta(b_1 \ldots b_q) \in \Delta^{(q)}$ is a full fundamental interval, then $r_\kappa = q$. Put $r_0 = 1$, then we can divide the block of digits $b_1 \ldots b_q$ into κ subblocks $C_1, \ldots, C_\kappa$, where $C_i = b_{r_{i-1}} \ldots b_{r_i - 1}$. So $\Delta(b_1 \ldots b_q) = \Delta(C_1 \ldots C_\kappa)$. These subblocks, C_i, have the following properties.

(i) If $|C_i|$ denotes the length of block C_i, then $|C_i| = r_i - r_{i-1}$ for all $i \in \{1, 2, \ldots, \kappa\}$.

(ii) If $b_{r_i} = 0$, then $r_i = r_{i-1} + 1$.

(iii) If $b_{r_i} = j \in \{2, 3\}$, then the block C_{i+1} is equal to j followed by the first part of the expansion of 1 if $j = 2$ and that of $1/\beta^3$ if $j = 3$. So $C_{i+1} = j d_1^{(j)} \ldots d_{|C_{i+1}|-1}^{(j)}$.

(iv) For all $i \in \{1, \ldots, \kappa\}$, $\Delta(C_i)$ is a full fundamental interval of rank $|C_i|$.

The above procedure gives for each full fundamental interval $\Delta(b_1 \ldots b_q)$, a subdivision of the block of digits $b_1 \ldots b_q$ into subblocks $C_1, \ldots, C_\kappa$, such that $\Delta(C_i)$ is a full fundamental interval of rank $|C_i|$ and $\Delta(b_1 \ldots b_q) = \Delta(C_1 \ldots C_\kappa)$. The next lemma is the last step in proving that the system $(R, \mathcal{B}, \nu, \mathcal{T})$ is a version of the natural extension

Lemma   3.1. *The σ-algebra $\mathcal{B}$ on R and   the σ-algebra $\bigvee_{n=0}^\infty \mathcal{T}^n \pi^{-1}(\mathcal{B}([0,2)))$ are equal.*

Proof. First notice that by Lemma 2.2, each of the σ-algebras $\mathcal{B}_{(j,n)}$ is generated by the direct products of the full fundamental intervals, contained in the rectangle $R_{(j,n)}$. Also, $\mathcal{B}_0$ is generated by the direct products of the full fundamental intervals. It is clear that $\bigvee_{n=0}^\infty \mathcal{T}^n \pi^{-1}(\mathcal{B}([0,2))) \subseteq \mathcal{B}$. For the other inclusion, first take a generating rectangle in R_0:

$$\Delta(a_1 \ldots a_p) \times \Delta(b_1 \ldots b_q) \times \{0\} \times \{0\},$$

where $\Delta(a_1 \ldots a_p)$ and $\Delta(b_1 \ldots b_q)$ are full fundamental intervals. For the set $\Delta(b_1 \ldots b_q)$ construct the subblocks $C_1, \ldots, C_\kappa$. By Lemma 2.1 $\Delta(C_\kappa C_{\kappa-1} \ldots C_1 a_1 \ldots a_p)$ is a full fundamental interval of rank $p+q$. Then

$$\mathcal{T}^q \left(\pi^{-1}(\Delta(C_\kappa C_{\kappa-1} \ldots C_1 a_1 \ldots a_p)) \cap R_0\right)$$
$$= \Delta(a_1 \ldots a_p) \times \Delta(b_1 \ldots b_q) \times \{0\} \times \{0\} \subseteq \bigvee_{n=0}^\infty \mathcal{T}^n \pi^{-1}(\mathcal{B}([0,2))).$$

Now let $\Delta(a_1 \ldots a_p) \times \Delta(b_1 \ldots b_q) \times \{j\} \times \{n\}$ be a generating rectangle for $\mathcal{B}_{(j,n)}$, for $j \in \{2, 3\}$ and $n \geq 1$. So $\Delta(a_1 \ldots a_p)$ and $\Delta(b_1 \ldots b_q)$ are again full fundamental intervals. Notice that

$$\Delta(b_1 \ldots b_q) \subseteq \Delta(\underbrace{0 \ldots 0}_{n \text{ times}}),$$

which means that $q \geq n$. Also $b_i = 0$ and thus $r_{i+1} = i + 1$ for all $i \in \{0, \ldots, n-1\}$. So, if we divide $b_0 \ldots b_{q-1}$ into subblocks C_i as before, we get that $C_1 = C_2 = \ldots = C_n = 0$, that $\kappa \geq n$ and that $|C_{n+1}| + \ldots + |C_\kappa| = q - n$. Consider the set $C = \Delta(C_\kappa C_{\kappa-1} \ldots C_{n+1} j d_1^{(j)} \ldots d_{n-1}^{(j)} a_1 \ldots a_p)$. This set is a fundamental interval of rank $p + q$ and $\mathcal{T}^q C = \Delta(a_1 \ldots a_p)$. Let $D = \pi^{-1}(C) \cap R_0$. Then $\mathcal{T}^q D = \Delta(a_1 \ldots a_p) \times \Delta(b_1 \ldots b_q) \times \{j\} \times \{n\}$. So,

$$\Delta(a_1 \ldots a_p) \times \Delta(b_1 \ldots b_q) \times \{j\} \times \{n\} \in \bigvee_{n=0}^\infty \mathcal{T}^n \pi^{-1}(\mathcal{B}([0,2))). \qquad \square$$

This leads to the following theorem, the proof of which follows from Remark 2.1, the properties of π and Lemma 3.1.

Theorem 3.1. *The dynamical system* $(R, \mathcal{B}, \nu, \mathcal{T})$ *is a version of the natural extension of the dynamical system* $([0, 2), \mathcal{B}([0, 2)), \mu, T)$, *where* $\mu = \nu \circ \pi^{-1}$ *is an invariant probability measure of* T, *equivalent to the Lebesgue measure on* $[0, 2)$, *whose density function,* $h : [0, 2) \rightarrow [0, 2)$, *is given by*

$$h(x) = \frac{1}{16 - 7\beta}[(1 + 2\beta)1_{[0,1/\beta^3)}(x) + (2 + \beta)1_{[1/\beta^3, 1/\beta^2)}(x)$$
$$+ 2\beta 1_{[1/\beta^2, 1/\beta)}(x) + \beta^2 1_{[1/\beta, 1)}(x) + \beta 1_{[1,\beta)}(x) + 1_{[\beta, 2)}(x)].$$

4. Towering the orbits

For the second version of the natural extension, we will define a transformation on a certain subset of $[0, 2) \times [0, 2\beta)$, using the transformation $\mathcal{T}$, defined in the previous section. Define for $n \geq 1$ the following intervals:

$$I_{(2,n)} = \left[\frac{2}{\beta^2} \sum_{j=0}^{n-1} \frac{1}{\beta^j}, \frac{2}{\beta^2} \sum_{j=0}^{n} \frac{1}{\beta^j} \right), \quad I_{(3,n)} = \left[2 + \frac{2}{\beta^2} \sum_{j=1}^{n-1} \frac{1}{\beta^j}, 2 + \frac{2}{\beta^2} \sum_{j=1}^{n} \frac{1}{\beta^j} \right),$$

where $\sum_{j=1}^{0} \beta^{-j} = 0$. Let $I_0 = [0, \frac{2}{\beta^2})$. Notice that all of these rectangles are disjoint and that $\bigcup_{n=1}^{\infty} I_{(2,n)} = [2\beta^{-2}, 2)$ and $\bigcup_{n=1}^{\infty} I_{(3,n)} = [2, 2\beta)$, so that these intervals together with I_0 form a partition of $[0, 2\beta)$. Now define the subset $I \subseteq [0, 2) \times [0, 2\beta)$ by

$$I = ([0, 2) \times I_0) \cup \bigcup_{n=1}^{\infty} (([0, T^{n-1}1) \times I_{(2,n)}) \cup ([0, T^{n-1}\frac{1}{\beta^3}) \times I_{(3,n)}))$$

and let the function $\phi : I \rightarrow R$ be given by

$$\phi(x, y) = \begin{cases} (x, \beta^2(y - \frac{2}{\beta^2} - \frac{2}{\beta^3} \sum_{j=0}^{n-1} \frac{1}{\beta^j}), 2, n), & \text{if } y \in I_{(2,n)}, \\ (x, \beta^2(y - 2 - \frac{2}{\beta^3} \sum_{j=0}^{n-1} \frac{1}{\beta^j}), 3, n), & \text{if } y \in I_{(3,n)}, \\ (x, \beta^2 y, 0, 0), & \text{if } y \in I_0. \end{cases}$$

So ϕ maps I_0 to R_0 and for all $n \geq 1$, $j \in \{2, 3\}$, ϕ maps $I_{(j,n)}$ to $R_{(j,n)}$. Clearly, ϕ is a measurable bijection. Define the transformation $\tilde{\mathcal{T}} : I \rightarrow I$, by $\tilde{\mathcal{T}}(x, y) = \phi^{-1}(\mathcal{T}(\phi(x, y)))$. It is straightforward to check that $\tilde{\mathcal{T}}$ is invertible. In Figure 4.1 we see this transformation.

Let $\mathcal{I}$ be the collection of Borel sets on I. If λ_2 is the 2-dimensional Lebesgue measure, then $\lambda_2(I) = 78 - 46\beta = \bar{\lambda}(R)/\beta^2$. Define a measure $\tilde{\nu}$ on $(I, \mathcal{I})$ by setting $\tilde{\nu}(E) = (\nu \circ \phi)(E)$, for all $E \in \mathcal{I}$. Then ϕ is measure preserving and the systems $(R, \mathcal{B}, \mu, \mathcal{T})$ and $(I, \mathcal{I}, \tilde{\nu}, \tilde{\mathcal{T}})$ are isomorphic. Notice that $\tilde{\nu}$ is the normalized 2-dimensional Lebesgue measure on $(I, \mathcal{I})$ and that the projection of $\tilde{\nu}$ on the first coordinate gives μ again. The following

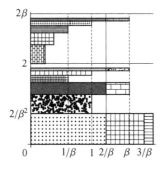 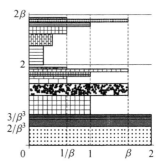

Fig. 4.1.   The transformation $\tilde{T}$ maps the regions on the left to the regions on the right.

lemma is now enough to show that $(I, \mathcal{I}, \tilde{\nu}, \tilde{T})$ is a version of the natural extension of $([0, 2), \mathcal{B}([0, 2)), \mu, T)$.

Lemma 4.1. *The σ-algebras $\mathcal{I}$ and $\bigvee_{n=0}^{\infty} \tilde{T}^n \pi^{-1}(\mathcal{B}([0, 2)))$ are equal.*

Proof. It is easy to see that $\bigvee_{n=0}^{\infty} \tilde{T}^n \pi^{-1}(\mathcal{B}([0, 2))) \subseteq \mathcal{I}$. For the other inclusion, notice that the direct products of full fundamental intervals contained in

$$([0, 2) \times I_0) \cup \bigcup_{n=1}^{\infty} ([0, T^{n-1}1) \times I_{(2,n)}),$$

generate the restriction of $\mathcal{I}$ to this set. If $\Delta(b_0 \ldots b_{n-1}) \in \Delta^{(n)}$ is full in $[0, \frac{2}{\beta})$, then the set $2 + \Delta(b_0 \ldots b_{n-1})$ is a subset of $[2, 2\beta)$. So the direct products of full fundamental intervals in $[0, \beta)$ and sets of the form $2 + \Delta(b_0 \ldots b_{n-1})$ contained in $\bigcup_{n=1}^{\infty} ([0, T^{n-1} \frac{1}{\beta^3}) \times I_{(3,n)})$, generate the restriction of $\mathcal{I}$ to this set. Since $\tilde{T}$ is isomorphic to T, the fact that

$$\mathcal{I} \subseteq \bigvee_{n=0}^{\infty} \tilde{T}^n \pi^{-1}(\mathcal{B}([0, 2)))$$

now can be proven in a way similar to the proof of Lemma 3.1.   □

References

1. A. Rényi, *Acta Math. Acad. Sci. Hungar* **8**, 477 (1957).
2. K. Dajani and C. Kalle, *Disc. and Cont. Dyn. Sys.* **18**, 199 (2007).
3. M. Pedicini, *Theoret. Comput. Sci.* **332**, 313 (2005).
4. K. Dajani and C. Kalle, To appear in *SMF Séminaires et Congrés*, (2009).
5. V. A. Rohlin, *Izv. Akad. Nauk SSSR Ser. Mat.* **25**, 499 (1961).

6. I. P. Cornfeld, S. V. Fomin and Y. G. Sinai, *Ergodic Theory*, Grundlehren der Mathematischen Wissenschaften [Fundamental Principles of Mathematical Sciences], Vol. 245 (Springer-Verlag, New York, 1982).
7. K. Dajani, C. Kraaikamp and B. Solomyak, *Acta Math. Hungar.* **73**, 97 (1996).
8. G. Brown and Q. Yin, *Ergodic Theory Dynam. Systems* **20**, 1271 (2000).

AN EQUIVARIANT TIETZE EXTENSION THEOREM FOR PROPER ACTIONS OF LOCALLY COMPACT GROUPS*

AASA FERAGEN

Department of Mathematics and Statistics
University of Helsinki
Finland

The classical Tietze extension theorem asserts that any continuous map $f: A \to \mathbb{R}^n$ from a closed subset A of a normal space X admits a continuous extension $F: X \to \mathbb{R}^n$. The Tietze-Gleason theorem is an equivariant version of the same theorem for spaces with actions of compact groups, proved by A. Gleason in the 1950s. Here we prove the following version of the theorem for proper actions of locally compact groups: let G be a locally compact group acting properly on a completely normal space X such that X/G is paracompact, and let A be a closed G-invariant subset of X. Suppose $\rho: G \to GL(n, \mathbb{R})$ is a representation; now any continuous G-equivariant map $f: A \to \mathbb{R}^n(\rho)$ admits a continuous G-equivariant extension $F: X \to \mathbb{R}^n(\rho)$.

Keywords: Proper actions; equivariant extension; Tietze.

1. Introduction

The Tietze extension theorem is one of the most basic, and perhaps the most well-known, continuous extension theorems. An equivariant version of it for compact groups was proven by A. Gleason in the 1950s [3], using the Haar integral to "average" over the group. The same technique does not work for non-compact groups since the Haar integral does not generally converge over non-compact groups. However, for proper actions of locally compact groups, we can use slices to pass from a solution for a compact subgroup on a slice, to a solution for the whole group on the corresponding tubular neighborhood.

*The research leading to this article was financed by Helsingin Yliopiston Tiedesäätiö and the Magnus Ehrnrooth Foundation.

2. Prerequisites

Throughout the article, we define locally compact spaces to be Hausdorff.

Let G be a locally compact topological group, and let X be a topological space. An *action* of G on X is a continuous map $\Phi\colon G \times X \to X$ such that $\Phi(e, x) = x$ for all $x \in X$, where e is the neutral element of the group, and such that $\Phi(g', \Phi(g, x)) = \Phi(g'g, x)$. We usually denote $\Phi(g, x) = gx$.

A topological space with a G-action on it is abbreviated as a G-space, and a subset $A \subset X$ is called a G-*subset* if $GA = A$. Similarly we say that a covering of a G-space is a G-covering if consists of G-sets, and a refinement by G-sets called a G-refinement.

We reserve the word *map* for continuous mappings. A map $f\colon X \to Y$ between G-spaces X and Y is G-equivariant if $f(gx) = gf(x)$ for all $g \in G$ and $x \in X$, and for short, such maps are called G-*maps*.

Two subsets A and B of a topological space X are said to be *separated* in case $\bar{A} \cap B = \emptyset = A \cap \bar{B}$. A topological space X is *completely normal* if any two separated subsets A and B of X are contained in disjoint neighborhoods V_A and V_B. A topological space X is completely normal if and only if any subspace of X is normal [4, Theorem II.5.1].

The action of G on a completely regular space X is a *Cartan action* if, for any $x \in X$, there exists a neighborhood V of x in X such that the set $\{g \in G | gV \cap V \neq \emptyset\}$ is relatively compact in G. The action is *proper* if for any pair of points $x, y \in X$ we can find neighborhoods V_x and V_y of x and y respectively, such that the set $\{g \in G | gV_x \cap V_y \neq \emptyset\}$ is relatively compact in G.

Obviously, any proper action is a Cartan action.

Let H be a closed subgroup of G. A subset S of a G-space X is an H-*slice* if GS is open in X and there exists a G-map $f\colon GS \to G/H$ such that $S = f^{-1}(eH)$.

Lemma 2.1. *Suppose that $S \subset X$ is an H-slice. Then*

i) $gS \cap S \neq \emptyset$ *implies that* $g \in H$, *and*
ii) S *is closed.*

Proof. To prove i), suppose that $gS \cap S \neq \emptyset$, and let $f\colon GS \to G/H$ be such that $S = f^{-1}(eH)$. Then for some $s \in S$, also $gs \in S$ and thus $gH = g(eH) = gf(s) = f(gs) \in f(S) = eH$. But this implies $g \in H$.

Claim ii) is trivial since f is continuous and eH is a closed point in G/H, since H is closed in G. $\qquad\square$

Given a closed subgroup H of G and an H-space S, there is an action of H on the product $G \times S$ given by $h(g, s) = (gh^{-1}, hs)$. We denote by $G \times_H S$ the quotient space $(G \times S)/H$, which is called the *twisted product* of G and S with respect to H. There is an action of G on $G \times_H S$ defined by the formula $\bar{g}[g, s] = [\bar{g}g, s]$.

Proposition 2.1. *Let H be a closed subgroup of G and let S be an H-invariant subset of X. Then S is an H-slice if and only if GS is an open subset of X and*

$$G \times_H S \approx_G GS.$$

Proof. See [1, Proposition 3.2]. □

We say that the open set GS is a *tubular neighborhood* (of x) if S is an H-slice (a slice at x). A *tubular covering* of a G-space X is a covering of X by tubular neighborhoods.

Lemma 2.2. *Suppose that $x \in X$, that GS is a tubular neighborhood of x in G, and that U is a G-invariant open neighborhood of x in GS. Then U can be given the structure of a tubular neighborhood as well.*

Proof. Let $f \colon GS \to G/H$ be a G-map such that $f^{-1}(eH) = S$, then $f| \colon U \to G/H$ is a G-map and $f|^{-1}(eH) = S \cap U =: S'$, making S' a slice. Then $U = GS'$ is the corresponding tubular neighborhood. □

There are many theorems on existence of tubular coverings, the original ones due to Palais; however, we shall use a very general version by H. Biller.

Theorem 2.1. *[2, Theorem 2.5] A continuous action of a locally compact group G on a completely regular space X is a Cartan action if and only if X admits a tubular covering, where each tubular neighborhood corresponds to an H-slice for some compact subgroup H of G.*

3. The equivariant Tietze extension theorem

Theorem 3.1. *Let G be a locally compact group and let X be a completely normal proper G-space such that X/G is paracompact. Let $A \subset X$ be a closed G-subset. Let $\rho \colon G \to \mathrm{GL}(n, \mathbb{R})$ be a representation and let $f \colon A \to \mathbb{R}^n(\rho)$ be a G-map. Then there exists a continuous G-extension $F \colon X \to \mathbb{R}^n(\rho)$.*

Before proving the theorem, we need a couple of short lemmas.

Lemma 3.1. *Let G be a locally compact group, let X and Y be G-spaces, let S be an H-slice for a compact subgroup H of G, and let V be an H-invariant subset of S. Let $f\colon V \to Y$ be an H-map. Then there exists a unique continuous G-map $\tilde{f}\colon GV \to Y$ which extends f.*

Proof. The only possible G-function $GV \to Y$ which extends f is $\tilde{f}$ defined by

$$\tilde{f}(gv) = gf(v) \ \forall g \in G, \ v \in V.$$

We show that the function $\tilde{f}$ is well-defined. If $gv = g'v'$ for some $g, g' \in G$ and $v, v' \in V \subset S$ then $v = g^{-1}g'v'$, implying $S \cap (g^{-1}g')S \neq \emptyset$. Since S is an H-slice, we get by Lemma 2.1 that $g^{-1}g' \in H$. Since f is an H-map, then $f(v) = f(g^{-1}g'v') = g^{-1}g'f(v')$ giving $gf(v) = g'f(v')$. Hence $\tilde{f}(gv) = \tilde{f}(g'v')$ and $\tilde{f}$ is well defined.

Equivariance is clear from the definition, and $\tilde{f}$ is continuous since the diagram below commutes and the left vertical projection is a quotient map because it is the projection $G \times V \to (G \times V)/H \approx GV$ to the orbit space of a compact group action.

$$
\begin{array}{ccc}
(g,v) & G \times V \xrightarrow{\ \mathrm{id}\times f\ } G \times Y \\
\Big\downarrow & \text{quotient}\Big\downarrow \qquad\qquad \Big\downarrow \text{action} \\
gv & GV \xrightarrow[\ \tilde{f}\ ]{} Y
\end{array}
\qquad \square
$$

Lemma 3.2. *Suppose that X admits a global slice; then the statement of Theorem 3.1 holds for X.*

Proof. Assume that $X = GS$ where S is an H-slice for a compact subgroup H of G. By [3] the map

$$\tilde{f}\colon A \cap S \to \mathbb{R}^n(\rho|),$$

where $\rho|\colon H \to GL(n, \mathbb{R})$, admits an H-equivariant extension

$$\tilde{F}\colon S \to \mathbb{R}^n(\rho|),$$

and by Lemma 3.1 there exists a G-equivariant continuous extension

$$F\colon GS \to \mathbb{R}^n(\rho)$$

given by $F(gs) = g\tilde{F}(s)$, where we easily see that $F|A = f$. $\qquad \square$

Proof of Theorem 3.1. By Theorem 2.1, any Cartan G-space has a tubular covering; hence, in particular, our G-space X admits a tubular covering $\{GS_i\}_{i\in I}$ where each S_i is an H_i-slice for some compact subgroup H_i of G.

Since X/G is paracompact, the covering $\{GS_i\}_{i\in I}$ has a locally finite open G-refinement $\{U'_j\}_{j\in J}$. Because Hausdorff paracompact spaces are normal and $\{U'_j\}_{j\in J}$ is, in particular, point-finite, we may find a G-refinement $\{U_j\}_{j\in J}$ of $\{U'_j\}_{j\in J}$ such that $\bar{U}_j \subset U'_j$ for all $j \in J$ by [4, Chapter II, Thm 4.2]. ($\bar{U}_j = \mathrm{cl}_X U_j$.) Now the coverings $\{U_j\}_{j\in J}$ and $\{\bar{U}_j\}_{j\in J}$ are locally finite as well.

Well-order the index set J, and let $j \in J$. Suppose that for all $k < j$ we have G-extensions $F_k \colon \bar{U}_k \to Y$ of $f|A \cap \bar{U}_k$, such that all F_k agree on their common domain. We claim that there exists a G-extension $F_j \colon \bar{U}_j \to Y$ of $f|A \cap \bar{U}_j$, which agrees with every F_k, $k < j$, on their common domain.

The maps F_k, for $k < j$, combine to a function $\bar{F}_j \colon \bigcup_{k<j} \bar{U}_k \to Y$, given by $\bar{F}_j|\bar{U}_k = F_k$. This function $\bar{F}_j$ is continuous because the family $\{\bar{U}_k\}_{k<j}$ is locally finite. For the same reason, the union $\bigcup_{k<j} \bar{U}_k \cup A$ is a closed subset of X.

By Lemma 3.2 the map $\bar{F}_j|\bar{U}_j \cap (\bigcup_{k<j} \bar{U}_k \cup A)$ has a G-extension $F_j \colon \bar{U}_j \to Y$; obviously F_j agrees with each F_k $(k < j)$ on their common domain.

Hence, by transfinite construction, there exist, for every $j \in J$, a G-extension $F_j \colon \bar{U}_j \to Y$ of $f|A \cap \bar{U}_j$ such that F_j agrees with any F_k on their common domain.

Using the above we obtain a G-map $F \colon X \to Y$ by setting

$$F(x) = F_j(x) \text{ when } x \in \bar{U}_j.$$

The map is well-defined, and it is continuous by local finiteness and closedness of the $\bar{U}_j$. Thus F is the wanted extension. $\square$

References

1. H. Biller, *Proper actions on cohomology manifolds*, Trans. AMS 355 No. 1, 2002, 407–432.
2. H. Biller, *Characterizations of proper actions*, Math. Proc. Camb. Phil. Soc. 136, 2004, 429–439.
3. A. Gleason, *Spaces with a compact Lie group of transformations*, Proc. Amer. Math. Soc, 1, 1950, 35–43.
4. S. Gaal, *Point set topology*, 1964.

ON UNIFORM TANGENTIAL APPROXIMATION BY LACUNARY POWER SERIES

GOHAR HARUTYUNYAN

Institut für Mathematik
Carl von Ossietzky Universität Oldenburg
26111 Oldenburg, Germany

Let $\Omega \subset \mathbb{C}$ be a simple connected domain with $0 \in \Omega$, $E \subset \Omega$ a set of tangential approximation with interior points. We consider the following problem: under what assumptions about E can each function continuous on E and holomorphic in E° be tangential approximated on E by functions holomorphic in Ω and having power series at 0 with caps of density 0. Some sufficient conditions are obtained.

Keywords: Uniform tangential approximation; lacunary approximation.

Notation and Introduction

For an arbitrary set $E \subset \mathbb{C}$ we denote by $E^\circ, \partial E, \overline{E}$ and E^c the interior, boundary, closure and complement of E in $\mathbb{C}$, respectively. Let $C(E)$ be the set of all functions continuous on E and $A(E)$ the subspace of $C(E)$ consisting of the functions holomorphic in E°. Finally, for a domain $\Omega \subset \mathbb{C}$ we denote by $H(\Omega)$ the set of all functions holomorphic in Ω.

Let $\Omega \subset \mathbb{C}$ be a domain and $E \subset \Omega$ a relatively closed subset of Ω. The problem of describing the relatively closed sets $E \subset \Omega$ of uniform and tangential approximation by functions holomorphic in Ω are studied in papers [1], [2] and [10], respectively.

In connection with this it is natural to consider the following question: Under what assumptions on E and a subsequence Q of natural numbers can each function $f \in A(E)$ be tangential approximated by functions $g \in H(\Omega)$ having expansions

$$g(z) = \sum_{n00}^{\infty} g_n z^n \quad \text{with} \quad g_n = 0 \quad \text{for} \quad n \notin Q \cup \{0\}?$$

The paper consists of two parts. In Section 1 we define the sets of uniform and tangential approximation and give their descriptions using the corresponding results from [1], [2] and [10]. In Section 2 we consider lacunary approximation; more precisely, Section 2.1 is about a result on uniform approximation by polynomials with caps of density 0 (more details in [8]), in Section 2.2 we prove an auxiliary proposition and in Section 2.3 we give and prove the main result (more details in [6]).

1. Uniform and tangential approximation by holomorphic functions

Let $\Omega \subset \mathbb{C}$ be a domain and $E \subset \Omega$ a relatively closed subset in Ω.

Definition 1.1.

(1) E is called a set of uniform approximation, if for any $f \in A(E)$ and any $\varepsilon > 0$ there exists an $g \in H(\Omega)$ such that

$$|f(z) - g(z)| < \varepsilon, \quad z \in E.$$

(2) E is called a set of tangential approximation, if for any $f \in A(E)$ and any $\varepsilon \in C(E), \varepsilon > 0$, there exists an $g \in H(\Omega)$ such that

$$|f(z) - g(z)| < \varepsilon(z), \quad z \in E.$$

Remark 1.1.

(1) Any set of tangential approximation is a set of uniform approximation.
(2) Let $E \subset \Omega$ be compact. Then E is a set of tangential approximation if and only if E is a set of uniform approximation.
(3) Without loss of generality by tangential approximation we can assume that

$$\lim_{\substack{z \to \zeta \\ \zeta \in \partial\Omega}} \varepsilon(z) = 0.$$

(4) Let $E^\circ = \emptyset$. Then E is a set of tangential approximation if and only if E is a set of uniform approximation.

The assertions i)-iii) are trivial. To show iv) let us first prove the following Lemma ([4], Ch. IV, 3, Lemma 3).

Lemma 1.1. *Suppose $E \subset \Omega$ is a set of uniform approximation and $\psi \in A(E)$. Then for every function $f \in A(E)$ and for $\varepsilon(z) = |e^{\psi(z)}|$ there is an $g \in H(\Omega)$ such that*

$$|f(z) - g(z)| < \varepsilon(z), \quad z \in E.$$

Proof. There is an $\tilde{g} \in H(\Omega)$ such that

$$|\psi(z) - \tilde{g}(z)| < 1, \quad z \in E.$$

Let $h(z) := e^{\tilde{g}(z)-1}$ and consider the function $\frac{f}{h} \in A(E)$. There is an $\tilde{\tilde{g}} \in H(\Omega)$ such that

$$\left| \left(\frac{f}{h} \right)(z) - \tilde{\tilde{g}}(z) \right| < 1, \quad z \in E.$$

Set $g(z) := h(z)\tilde{\tilde{g}}(z)$. It follows that

$$|f(z) - g(z)| < |h(z)| = \exp\{\operatorname{Re} \tilde{g}(z) - 1\}$$
$$< \exp\{\operatorname{Re} \psi(z)\} = |e^{\psi(z)}|, \quad z \in E,$$

which completes the proof. □

Let now $E \subset \Omega, E^\circ = \emptyset$, be a set of uniform approximation and $\varepsilon \in C(E), \varepsilon > 0$. Then $\log \varepsilon \in A(E)$ and by Lemma 1.1 for any $f \in A(E)$ there is an $g \in H(\Omega)$ such that

$$|f(z) - g(z)| < e^{\log \varepsilon(z)} = \varepsilon(z).$$

This proves iv) of Remark 1.1. □

According to Remark 1.1 we see a difference between two kinds of approximation only in the case if E is an "exact relative" closed subset of Ω and $\varepsilon(z) \to 0$ as $z \to \zeta \in \partial\Omega$.

The **first example of a set of tangential approximation**: $\mathbb{R}$ is a set of tangential approximation in $\mathbb{C}$ (see [3]).

The next point of this section is to describe the sets of uniform/tangential approximation.

We denote the one-point compactification of Ω by $\Omega^* = \Omega \cap \{\infty\}$.

Definition 1.2.

(1) *(K-condition)* We say that $E \subset \Omega$ satisfies the K-condition, if for any neighbourhood U of ∞ in Ω there is a neighbourhood $\tilde{U} \subset U$ of ∞ such that any $z \in E^c \cap \tilde{U}$ can be connected to ∞ by an arc γ_z in $E^c \cap U$.

(2) *(A-condition)* We say that $E \subset \Omega$ satisfies the A-condition, if for any neighbourhood U of ∞ in Ω there is a neighbourhood $\tilde{U} \subset U$ of ∞ such that no component of E° meets both $\tilde{U}$ and $\Omega^* \setminus U$.

Theorem 1.1.

(1) (Arakelian [1] and [2]) *A (relative) closed subset $E \subset \Omega$ is a set of uniform approximation if and only if E satisfies the K-condition.*

(2) (Nersesian [10]) *A (relative) closed subset $E \subset \Omega$ is a set of tangen-tial approximation if and only if E satisfies the K-condition and A-condition.*

The first result about the A-condition is given by P. Gauthier (see [5]).

2. Lacunary approximation

2.1. *Uniform approximation by lacunary polynomials*

The classical result of Mergelyan, proved in 1951, solved the problem about approximation by polynomials: the system of functions $\{z^n\}_{n=0}^\infty$ is complete in $A(E)$ if and only if E^c is connected (see [5], Ch. II, 2, Theorem 1). The natural question is: under what assumptions about compact E and subse-quence $\{p_n\}_{n=0}^\infty, p_0 = 0$, the system of powers in Theorem of Mergelyan can be replaced by the system of functions $\{z^{p_n}\}_{n=0}^\infty$, i.e., when any function $f \in A(E)$ can be uniformly approximated on E by polynomials

$$p(z) = \sum_{n=0}^m c_n z^{p_n}?$$

It is obvious that if $\{p_n\}_{n=1}^\infty \neq \mathbb{N}$ the position of 0 with respect to the interior of E is essential, i.e., in general the condition $0 \notin E^\circ$ is necessary. The case of $0 \in E^\circ$ is studied by J. Korevaar in [7]. Here we assume that $0 \notin E^\circ$.

Theorem 2.1. *(Martirosian [8]) Let $E \subset \mathbb{C}$ be a compact and $\{p_n\}_{n=1}^\infty \subset \mathbb{N}$ a subsequence with the following properties*

(1) i1) E^c *is connected;*
    i2) $0 \notin E^\circ$;
    i3) $0 \notin \partial E_n$ *for* $n = 1, 2, \ldots$, *where* $\{E_n\}_{n=1}^\infty$ *are the components of* E°;
(2) $\lim \frac{n}{p_n} = 1$.

Then the system of functions

$$\{z^{p_n}\}_{n=0}^\infty, p_0 = 0, \tag{1}$$

is complete in $A(E)$.

Concerning the strength of the assumptions note that i1) and i2) are necessary if we (in general) consider lacunary approximation by polynomi-als. The next theorem shows the essentiality of the condition i3).

Theorem 2.2. [8] *Let* $\{p_n\}_{n=1}^{\infty} \subset \mathbb{N}$ *be a subsequence such that*

$$\sum_{n=1}^{\infty} \frac{1}{q_n} = \infty$$

with $\{q_n\}_{n=1}^{\infty} := \mathbb{N} \setminus \{p_n\}_{n=1}^{\infty}$. *Then there is a compact* $E \subset \mathbb{C}$ *with connected complement* $E^c, 0 \in \partial E, E = \overline{E^{\circ}}$ *and such that the system* (1) *is no complete in* $A(E)$.

Since a subsequence $\{q_n\}_{n=1}^{\infty}$ can satisfy the both conditions

$$\lim_{n \to \infty} \frac{n}{q_n} = 0, \quad \sum_{n=1}^{\infty} \frac{1}{q_n} = \infty$$

(for example $q_n = n[\log n]$), from Theorem 2.2 (in general) follows the condition i3).

Finally, we give an example to show the necessity of the condition ii).

Example 2.1. Set

$$E = \left(\bigcup_{q=2}^{\infty} E_q \right) \bigcup \{0\} \quad \text{with} \quad E_q = \{z \in \mathbb{C} : 2^{q^2} z^q = 1\}.$$

It is not difficult to prove that for any $\delta, 0 < \delta < 1$, there is a subsequence $\{p_n\}_{n=1}^{\infty} \subset \mathbb{N}$ with density δ and such that the system $\{z^{p_n}\}_{n=1}^{\infty} \cup \{1\}$ is no complete in $A(E)$. More details see in [8].

2.2. *Auxiliary Proposition*

We set

$$D_r := \{z : |z| < r\} \quad \text{for} \quad 0 < r < \infty; \quad D_0 = \emptyset.$$

Let μ be a complex Borel measure on the compact $E \subset \mathbb{C}$; by $|\mu|$ we denote its variation. We denote by G the Cauchy transform

$$G(t) = \int_E \frac{d\mu(z)}{t - z}, \quad t \in \overline{\mathbb{C}} \setminus E.$$

We have $G \in H(\overline{\mathbb{C}} \setminus E)$ and $G(\infty) = 0$.

Lemma 2.1. *Let* $E \subset \mathbb{C}$ *be a compact set with connected complement* E^c, D_r *the maximal disk containing in* E°, μ *a complex Borel measure on* ∂E *and let* G *be its Cauchy transform. If there exists a function* $G_1 \in H(\overline{\mathbb{C}} \setminus \overline{D}_r)$ *such that*

$$G(t) = G_1(t) \quad \text{for} \quad t \in \overline{\mathbb{C}} \setminus E, \tag{2}$$

then the equation (2) *holds for every* $t \in \partial E \setminus \overline{D}_r$ *for which converges the integral*

$$\int\limits_{\partial E} \frac{d|\mu|(z)}{|t - z|}.$$

In the following lemma $E \subset \mathbb{C}$ is a compact set with connected complement E^c, D_r the maximal disk contained in E°, c any component of the interior E° for which $\partial c \cap \partial D_r = \emptyset$ and μ a complex Borel measure concentrated on ∂E. Let G be the Cauchy transform of μ. We decompose μ into a sum of measures $\mu = h_c + \sigma_c$, where h_c is absolutely continuous and σ_c singular with respect to every harmonic measure λ_a defined on ∂E for $a \in c$.

Lemma 2.2. *Suppose that the Cauchy transform of* μ *satisfies* (2) *where* $G_1 \in H(\overline{\mathbb{C}} \setminus \overline{D}_r)$. *Then*

$$\int\limits_{\partial E} \frac{d\sigma_c(z)}{t - z} = G_1(t) \quad for \quad t \in (\overline{\mathbb{C}} \setminus E) \cup c. \tag{3}$$

We say that a compact set $E \subset \mathbb{C}$ satisfies condition (C) if

(1) the complement E^c of E is connected;
(2) 0 belongs to the interior E° of E. By E_i we denote the components of E° and assume that $0 \in E_1$;
(3) ∂E_1 is a smooth curve and $\overline{E}_i \cap \overline{D}_r = \emptyset$ for $i = 2, 3, \ldots$, where D_r is the maximal disk contained in E_1

Proposition 2.1. *Let* $\{q_n\}_{n=1}^\infty = Q \subset \mathbb{N}$ *be a subsequence with density* 1 *and a compact set* $E \subset \mathbb{C}$ *satisfies* (C). *Then every function* $f \in A(E)$ *having a representation*

$$f(z) = \sum_{n=0}^{\infty} f_n z^n, \quad f_n = 0 \quad for \quad n \notin Q \cup \{0\} \tag{4}$$

can be approximated uniformly on E *by polynomials of the form*

$$p(z) = \sum_{n=0}^{m} p_n z^n, \quad p_n = 0 \quad for \quad n \notin Q \cup \{0\}.$$

Proof. Let for an arbitrary complex Borel measure μ on ∂E the relations

$$\int\limits_{\partial E} z^{q_n} d\mu(z) = 0 \quad for \quad n = 0, 1, \ldots (q_0 = 0) \tag{5}$$

are satisfied. By virtue of Hahn-Banach and Riesz theorems it is sufficient to prove that any function $f \in A(E)$ possessing a representation (4) satisfies the condition

$$\int_{\partial E} f(z) d\mu(z) = 0. \tag{6}$$

For each fixed $t \in E^c$ with $|t|$ sufficiently large the power series

$$\sum_{n=0}^{\infty} \left(\frac{z}{t}\right)^n = \frac{1}{1 - \frac{z}{t}}$$

converges uniformly for $z \in \partial E$. Integrating this series with respect to μ and using (5) we get

$$G(t) = \sum_{n=1}^{\infty} t^{-p_n - 1} \int_{\partial E} z^{p_n} d\mu(z),$$

where $\{p_n\}_{n=1}^{\infty} = \mathbb{N} \setminus Q$. Since $G \in H(\overline{C} \setminus E)$, E^c is connected and $\{p_n\}_{n=1}^{\infty}$ has zero density, by Fabry's theorem the series of powers of $\frac{1}{t}$ in the above equation defines a function $G_1 \in H(\overline{C} \setminus \overline{D}_r)$. Consequently by the uniqueness theorem for analytic functions we obtain (2).

Now we decompose $\mu = h_{E_1} + \sigma_{E_1}$, where h_{E_1} is absolutely continuous and σ_{E_1} is singular with respect to any harmonic measure on ∂E; we take $a \in E_1$. By Lemma 2.2 we have

$$\int_{\partial E} z^n dh_{E_1(z)} = 0 \quad \text{for} \quad n = 0, 1, \ldots$$

and

$$\int_{\partial E} \frac{d\sigma_{E_1}(z)}{t - z} = G_1(t) \quad \text{for} \quad t \in \overline{\mathbb{C}} \setminus E, t \in E_1.$$

Similarly, $\sigma_{E_1} = h_{E_2} + \sigma_{E_2}$, where h_{E_2} is absolutely continuous and σ_{E_2} is singular with respect to any harmonic measure on ∂E; we take $a \in E_2$. Again by Lemma 2.2 we have

$$\int_{\partial E} z^n dh_{E_2}(z) = 0 \quad \text{for} \quad n = 0, 1, \ldots$$

and

$$\int_{\partial E} \frac{d\sigma_{E_2}(z)}{t - z} = G_1(t) \quad \text{for} \quad t \in \overline{\mathbb{C}} \setminus E, t \in E_1 \cup E_2.$$

Continuing this process we finally obtain:

$$\mu = \sum_{k=1}^{\infty} h_{E_k} + \sigma, \tag{7}$$

where the series of measures converges in the total variation norm, and we have

$$\int_{\partial E} z^n dh_{E_k}(z) = 0 \quad \text{for} \quad n = 0, 1, \ldots, k = 1, 2, \ldots \tag{8}$$

and

$$\int_{\partial E} \frac{d\sigma(z)}{t - z} = G_1(t) \quad \text{for} \quad t \in \overline{\mathbb{C}} \setminus E, t \in \bigcup_{k=1}^{\infty} E_k. \tag{9}$$

By Mergelyan's Theorem it follows from (8) that

$$\int_{\partial E} f(z) dh_{E_k}(z) = 0 \quad \text{for} \quad k = 1, 2, \ldots. \tag{10}$$

Further from (5) and (8) we find

$$\int_{\partial E} z^{q_n} d\sigma(z) = 0 \quad \text{for} \quad n = 0, 1, \ldots.$$

By virtue of Lemma 2.1, Lemmas 1,4 from [8] and (9) it follows that σ is concentrated on ∂E_1. Therefore

$$\int_{\partial E_1} z^{q_n} d\sigma(z) = 0 \quad \text{for} \quad n = 0, 1, \ldots.$$

However, since ∂E_1 is a smooth curve and the sequence Q is of density 1, applying the result of Korevaar and Dixon [7] we obtain:

$$\int_{\partial E_1} f(z) d\sigma(z) = 0.$$

Now (6) follows from (7) and (10). This completes the proof of the proposition. $\qquad\square$

2.3. *The main result*

Theorem 2.3. Harutyunyan and Martirosian [6] *Let $Q = \{q_n\} \subset \mathbb{N}$ be a subsequence with density 1, $\lim\limits_{n \to \infty} \frac{n}{q_n} = 1$, $\Omega \subset \mathbb{C}$ be a simply connected domain with $0 \in \Omega$ and let $E \subset \Omega$ be a set of tangential approximation, satisfying the conditions:*

(1) $0 \notin E^\circ$ and $0 \notin \partial E_n$ for $n = 1, 2, \ldots$, where $\{E_n\}_{n=1}^\infty$ are the components of E°;

(2) for every compact $F \subset \Omega$ the set of indices n for which $F \cap E_n \neq \emptyset$ is finite.

Then for arbitrary functions $f \in A(E)$ and $\varepsilon \in C(E), \varepsilon > 0$, there exists a $g \in H(\Omega)$ having an expansion

$$g(z) = \sum_{n=0}^\infty g_n z^n, \quad g_n = 0 \quad for \quad n \notin Q \cup \{0\} \tag{11}$$

and such that

$$|f(z) - g(z)| < \varepsilon(z), \quad z \in E. \tag{12}$$

Proof. We denote by $\rho(z, \partial\Omega)$ the spherical distance between $z \in \Omega$ and $\partial\Omega$ and

$$V_r(\partial\Omega) := \{z \in \Omega : \rho(z, \partial\Omega) < r\}, \quad 0 < r < \infty.$$

As noted in Section 1, without loss of generality we may suppose that the function ε depends solely on $\rho(z, \partial\Omega)$ and decreases to zero as $\rho(z, \partial\Omega) \to 0$. Since E is a set of tangential approximation, to each $r \in (0, \rho(0, \partial\Omega))$ corresponds an $r^* \in (0, r)$ such that any point $z \in E^c \cap V_{r^*}(\partial\Omega)$ can be joined to $\partial\Omega$ by a continuous curve $\gamma_z \subset E^c \cap V_r(\partial\Omega)$. From ii) it follows that no component of the set E° intersects both curves $\alpha_r := \{z \in \Omega : \rho(z, \partial\Omega) = r\}$ and $\alpha_{r^*} := \{z \in \Omega : \rho(z, \partial\Omega) = r^*\}$. For $r > 0$ we denote $A_r := \Omega \setminus V_r(\partial\Omega)$ and by B_r we denote the set of points in $E^c \cap A_r^c$ which can be joined to $\partial\Omega$ by continuous curves hitting A. We set $H_r := A_r \cup B_r$ and denote by $E_{i,r}, i = 1, 2, \ldots, s(r)$, all components of the set E° for which $\overline{E}_{i,r} \cap H_r \neq \emptyset$; by ii) they are finite in number. Note that the complement Ω_r^c of

$$\Omega_r := H_r \bigcup \left(\bigcup_{i=1}^{s(r)} \overline{E}_{i,r} \right)$$

is connected (see [5]).

We choose a sequence of positive numbers $\{r_n\}_{n=0}^\infty$ $(r_0 < \rho(0, \partial\Omega))$ decreasing to zero and such that $r_n = (r_{n-1}^*)^*, n = 1, 2, \ldots$. We set $\Omega_n := \Omega_{r_n}$ and

$$F_0 := E \cap \Omega_0, \quad F_n := \Omega_{n-1} \cup (E \cap (\Omega_n \setminus \Omega_{n-1})) \quad \text{for} \quad n = 1, 2, \ldots.$$

Evidently, $F_n \subset \mathbb{C}$ is compact and F_n^c connected; the latter follows from the equality

$$F_n^c = (\Omega_{n-1}^c \cap E^c) \cup \Omega_n^c, \quad n = 1, 2, \ldots.$$

We consider the set

$$\overline{E^\circ} \cap (\Omega_n \setminus \Omega_{n-1}). \tag{13}$$

Applying the reasoning used in [5] we obtain that each component K of the interior of (13) is a component of E° contained in Ω_n; the intersection $\overline{K} \cap \Omega_{n-1}$ contains not more than one point; the components K are finite in number. The components of the interior of (13) whose closures hit Ω_{n-1} we denote by $K_{j,n}, j = 1, 2, \ldots, l(n)$, and set $p_{j,n} := \overline{K}_{j,n} \cap \Omega_{n-1}$. Let us show that $p_{j,n} \notin \partial D_r$, where D_r is the aximal disk contained in Ω_{n-1}. Indeed, on one hand from the definition of Ω_{n-1} we have $\overline{K}_{j,n} \cap H_{r_n-1} = \emptyset, j = 1, 2, \ldots, l(n)$, and on the other, $\overline{K}_{j,n} \cap \left(\bigcup_{i=1}^{s(r_{n-1})} E_{i,r_{n-1}} \right) = \emptyset$. It remain to note that the definition of Ω_{n-1} and the condition ii) imply

$$\partial D_r \subset H_{r_{n-1}} \bigcup \left(\bigcup_{i=1}^{s(r_{n-1})} E_{i,r_{n-1}} \right).$$

Let us construct by induction a suitable sequence $\{p_j\}_{j=0}^\infty$ of polynomials of the form

$$p(z) = \sum_{n=0}^m p_n z^n, \quad p_n = 0 \quad \text{for} \quad n \notin Q \cup \{0\}. \tag{14}$$

We denote $t_n := \min\{\rho(z, \partial\Omega) : z \in \Omega_n\}$. We start by setting $f_0(z) = f(z)$ for $z \in F_0$. By Theorem 2.1 there is a polynomial p_0 of the form (14) satisfying

$$|f_0(z) - p_0(z)| < \frac{\varepsilon(t_2)}{2^5}, \quad z \in F_0.$$

Suppose that polynomials p_j of the form (14) have been constructed for $j = 0, 1, \ldots, n - 1$, satisfying

$$|f(z) - p_j(z)| < \begin{cases} \frac{\varepsilon(t_{j+1})}{2^{j+1}} & \text{for} \quad z \in E \cap (\Omega_j \setminus \Omega_{j-1}) \\ \frac{\varepsilon(t_{j+2})}{2^{j+4}} & \text{for} \quad z \in E \cap \partial\Omega_j \end{cases}$$

and

$$|p_j(z) - p_{j-1}(z)| < \frac{\varepsilon(t_{j+2})}{2^{j+5}}, \quad z \in \Omega_{j-1}.$$

Here we set $\Omega_{-1} = \emptyset$ and $p_{-1} = p_0$.

Now we define the polynomial p_n. First note that for the points $p_{j,n}, j = 1, 2, \ldots, l(n)$, there exists a simply connected domain $G_{n-1}(\overline{G}_{n-1} \subset \Omega)$ with smooth boundary ∂G_{n-1}, such that

$$\Omega_{n-1} \subset G_{n-1} \cup \{p_{j,n} : j = 1, 2, \ldots, l(n)\},$$

$$\overline{G}_{n-1} \cap \left(E \cap \left(\overline{\Omega_n \setminus \Omega_{n-1}} \right) \right) = \{p_{j,n} : j = 1, 2, \ldots, l(n)\}$$

and $p_{j,n} \notin \partial D_r$, where D_r is the maximal disk contained in G_{n-1}. We set $f_n(z) = p_{n-1}(z)$ for $z \in \overline{G}_{n-1}$. Applying the method used in [5] we extend the function f_n on $\Omega_n \setminus \Omega_{n-1}$ in a way to have $f_n \in A\big(\overline{G}_n \cup (E \cap (\Omega_n \setminus \Omega_{n-1}))\big)$ and

$$|f(z) - f_n(z)| < \begin{cases} \frac{\varepsilon(t_{n+1})}{2^{n+2}} & \text{for} \quad z \in E \cap \left(\overline{\Omega_n \setminus \Omega_{n-1}} \right) \\ \frac{\varepsilon(t_{n+2})}{2^{n+5}} & \text{for} \quad z \in E \cap \partial\Omega_n \end{cases}.$$

By Proposition 2.1 there is a polynomial p_n of the form (14) satisfying

$$|f_n(z) - p_n(z)| < \frac{\varepsilon(t_{n+2})}{2^{n+5}}, \quad z \in \overline{G}_n \cup (E \cap (\Omega_n \setminus \Omega_{n-1})).$$

Hence it follows

$$|f(z) - p_n(z)| < \begin{cases} \frac{\varepsilon(t_{n+1})}{2^{n+1}} & \text{for} \quad z \in E \cap (\Omega_n \setminus \Omega_{n-1}) \\ \frac{\varepsilon(t_{n+2})}{2^{n+4}} & \text{for} \quad z \in E \cap \partial\Omega_n \end{cases} \tag{15}$$

and

$$|p_n(z) - p_{n-1}(z)| < \frac{\varepsilon(t_{n+2})}{2^{n+5}}, \quad z \in \Omega_{n-1}. \tag{16}$$

The desired sequence of polynomials $\{p_n\}_{n=0}^{\infty}$ is constructed.

It follows from (16) that the sequence $\{p_n\}_{n=0}^{\infty}$ converges locally uniform in Ω to some function $g \in H(\Omega)$. Evidently in a neighbourhood of the origin g has the representation (11). If $z \in E$ then $z \in \Omega_m \setminus \Omega_{m-1}$ for some m. For every $n > m$ we have

$$|f(z) - g(z)| \le |f(z) - p_m(z)| + \sum_{j=m+1}^{n} |p_j(z) - p_{j-1}(z)| + |p_n(z) - g(z)|.$$

Therefore taking into account (15), (16) and choosing n sufficiently large we obtain (12). This completes the proof of theorem. $\qquad \square$

References

1. N.U. Arakelian, Uniform approximation on closed sets by entire functions. Izv. Akad. Nauk S.S.S.R., Ser. Mat., **28**, 5 (1964), 1187–1206.
2. N.U. Arakelian, Uniform and tangential approximation with analytic functions. Izv. Akad. Nauk Armenian SSR, Ser. Mat., **3**, 4-5 (1968), 273–286.
3. T. Carleman, "Sur un theoreme de Weierstrass", Ark. Mat. Astron. Fys., **20**, 4 (1927), 1–5.
4. D. Gaier, Lectures on complex approximation. Birkhäuser, 1985.
5. P. Gauthier, Tangential approximation by entire functions and functions holomorphic in a disc. Izv. Akad. Nauk Armenian SSR, Set. Mat., **4**, 5 (1969), 319–326.
6. G. Harutyunyan and V.A. Martirosian, On uniform tangential approximation by lacunary power series on carleman sets, II. Izv. Akad. Nauk Armenii. Matematika, **30**, 4 (1995), 68–75.
7. J. Korevaar, Lacunary forms of Walsh's approximation theorems. Theory of approximation of functions, Moscow, Nauka, 1977, 229–237.
8. V.A. Martirosian, On uniform complex approximation by lacunary polynomials. Math. USSR Sbornik, **48**, 2 (1984), 445–462.
9. V.A. Martirosian, On uniform tangential approximation by lacunary power series on carleman sets, I. Izv. Akad. Nauk Armenii. Matematika, **30**, 4 (1995), 58–67.
10. A.A. Nersesian, Carleman sets. Izv. Akad. Nauk Armenian SSR, Ser. Mat., **6** (1971), 465–471. MR **46**, 66. IV:3.

CYCLIC DIVISION ALGEBRAS IN SPACE-TIME CODING:
A BRIEF OVERVIEW

C. HOLLANTI

Department of Mathematics, FI-20014 University of Turku
Turku, Finland
cajoho@utu.fi
www.math.utu.fi

This paper aims at giving an introduction on space-time (ST) coding and on how cyclic division algebras (CDAs) fit into the picture. First, we describe the typical radio transmission model and then proceed to the notion of space-time coding and further on to multiple-input multiple-output (MIMO) channels. Cyclic division algebras provide us with a useful tool for designing codes for MIMO channels. The theory of their noncommutative orders allows us to build denser lattices, i.e. to pack more codewords in a given signal space. Finally, we establish a congenial connection between the density and the discriminant of an order. The theory is supported by illuminating examples. The material presented here is mainly reviewed from our (Vehkalahti-Hollanti-Lahtonen-Ranto) recent submission "On the densest MIMO lattices from cyclic division algebras" [1].

Keywords: Cyclic division algebras (CDAs); dense lattices; discriminants; maximal orders; multiple-input multiple-output (MIMO) channels; nonvanishing determinant (NVD).

1. Space-time coding: Idea and design criteria

Multiple-antenna wireless communication promises very high data rates, in particular when we have perfect channel state information (CSI) available at the receiver. In [2] the design criteria for such systems were developed, and further on the evolution of space-time (ST) codes took two directions: trellis codes and block codes. Our work concentrates on the latter branch.

Radio signals are typically modeled as vectors $x = (x_1, ..., x_n)$ with either complex or real components. In a classical radio channel the signals are spread in time only, and a distorted version $y = (y_1, ..., y_n) = hx + n$ of the transmitted vector x is received. Here h is the channel coefficient and n is the noise vector, both modeled as random variables subject to some statistics. Occasionally, h becomes very small and problems are met at the receiving end.

In order to fight fading, the notion of space-time coding was brought into the playground, and the very first space-time code called *Alamouti code* (see Example 1) was derived using the ring of Hamiltonian quaternions. Space-time coding seeks to add diversity by spreading the signal also in space, i.e. the transmission happens simultaneously from several spatially separated antennas. Signals (= codewords) now become matrices $X = (X_1 \cdots X_n)^T$ so that the ith antenna transmits the ith row X_i. The columns represent time; the jth column is transmitted in the jth time slot. The received signal is a linear combination $y = (h_1, ..., h_k)X + n$ of the signals transmitted by the various antennas, each multiplied by its own channel coefficient h_i. Assuming that the channel coefficients are independent random variables, it is very rare that they would all be small simultaneously.

Basic definitions and notions A *space-time (ST) code* $\mathcal{C}$ is a finite set of complex matrices of the same type. In our work, we only consider matrices coming from a lattice, as the discrete structure of a lattice will help us to meet the code design criteria presented below.

A *lattice* for us is a discrete finitely generated free abelian subgroup L of a real or complex finite dimensional vector space V, called the ambient space. In the space-time setting a natural ambient space is the space $\mathcal{M}_n(\mathbb{C})$ of complex $n \times n$ matrices. In order to transmit a maximal possible amount of information, we only consider full rank lattices that have a basis $x_1, x_2, \ldots, x_{2n^2}$ consisting of matrices that are linearly independent over the field of real numbers. It is well known that the measure, or hypervolume, $m(L)$ of the fundamental parallelotope of the lattice equals the square root of the determinant of the *Gram matrix* $G(L) = \left(\Re tr(x_i x_j^H)\right)_{1 \le i,j \le 2n^2}$, where ^{H} indicates the complex conjugate transpose of a matrix.

From the pairwise error probability (PEP) point of view [3], the performance of a space-time code is dependent on two parameters: *diversity gain* and *coding gain*. Diversity gain is the minimum of the rank of the difference matrix $X - X'$ taken over all distinct code matrices $X, X' \in \mathcal{C}$, also called the *rank* of the code $\mathcal{C}$. For non-zero square matrices, being full-rank coincides with being invertible. When $\mathcal{C}$ is full-rank, the coding gain is proportional to the determinant of the matrix $(X - X')(X - X')^H$. The minimum of this determinant taken over all distinct code matrices is called the *minimum determinant* of the code $\mathcal{C}$. If it is bounded away from zero even when the code size increases unlimitedly, the ST code is said to have the *nonvanishing determinant* (NVD) property.

One has to note that the determinant criterion is asymptotic in nature. For practical signal-to-noise ratios (SNR) also individual singular values play a role. Thus, simulations are necessary when deciding the better code.

The energy required for transmitting a certain code matrix equals the Frobenius norm of the matrix. Therefore if one wishes to save energy, dense lattices should be searched for. The natural task henceforth is to minimize $m(L)$ for a fixed determinant. E.g. to compare different lattices, we normalize them in such a way that the minimum determinant of each lattice is one, and only then compute $m(L)$. The smaller the measure the better the lattice for ST-coding purposes. Alternatively, we may compare the minimum determinants of different lattices after normalizing them all to have a unit measure. Once we have fixed the infinite lattice to be used, then in order to minimize the transmission energy, the finite code is formed by taking the desired number of matrices with the smallest Frobenius norms.

Forming such dense lattices with the NVD property is surprisingly difficult. It has turned out that tools from class field theory are needed in order to find the densest lattices. By using cyclic division algebras and their maximal orders, one has a complete control over the density of the resulting lattices.

Example 1. The Hamiltonian quaternions form a neat set for illustrating the above. Let $i^2 = j^2 = k^2 = -1$, and $ij = k$. If a, b, c, and d range over $\mathbb{R}$, we define the set $\mathbf{H}$ of Hamiltonian quaternions as the one containing the elements $q = a + bi + cj + dk$. This set becomes a ring by extending the above multiplication rules linearly. It might be helpful for the reader to note that $\mathbf{H} \simeq \mathbb{C} \oplus \mathbb{C}j$. The conjugate quaternion $\bar{q} = a - bi - cj - dk$ tells us that $q\bar{q} = a^2 + b^2 + c^2 + d^2 \in \mathbb{R} \setminus \{0\}$, whenever $q \neq 0$. Thus, the quaternions form a division algebra.

The quaternions can be conveniently represented either by complex 2×2-matrices or by real 4×4-matrices with respect to a suitable basis. Write now $z_1 = a + bi$ and $z_2 = c + di$. The left regular representation of the element $q = z_1 + jz_2$ with respect to the basis $\{1, j\}$ gives us (with the abuse of notation) $q = \begin{pmatrix} z_1 & -z_2^* \\ z_2 & z_1^* \end{pmatrix}$.

The aforementioned Alamouti code is obtained by selecting complex integer vectors (z_1, z_2) and mapping them to words of 2-antenna ST-code as above. The rank criterion is automatically met, and the minimum determinant of q is the squared minimum Euclidean distance.

In the above example, two gaussian symbols z_1 and z_2 are transmitted in two time slots, i.e. the *code rate* is one (symbols per time slot). Single antenna receiver can only deal with one complex information symbol (e.g. a gaussian integer) per unit of time, so in order to increase the code rate one needs to use more than one receiving antenna. In this *multiple-input multiple-output* (MIMO) setting the received signal becomes a matrix $Y_{n_r \times \ell} = H_{n_r \times n_t} X_{n_t \times \ell} + N_{n_r \times \ell}$, where n_t, n_r, and ℓ denote the number of transmitting antennas, receiving antennas, and the block length respectively.

2. Cyclic division algebras and orders

The NVD property can be achieved by using suitable orders of a cyclic division algebra. It has been shown that CDA based square ST codes with the NVD property achieve the diversity-multiplexing tradeoff (DMT) introduced by Zheng and Tse. This result raised an enormous amount of interest in cyclic division algebras. The DMT (again asymptotically) describes the maximum amount of information that can be transmitted at a given error rate, or similarly it can be thought of as the best possible performance a fixed size code can have. Do note that DMT is defined as a function of SNR, i.e. a family of codes (one for each SNR) is considered, not just a single code.

The theory of cyclic algebras and their representations as matrices are thoroughly considered in [4] and [5]. We are only going to recapitulate the essential facts here. A good introduction to class field theory is provided in [6].

In the following, we consider number field extensions E/F, where F denotes the base field and F^* (resp. E^*) denotes the set of the non-zero elements of F (resp. E). The rings of algebraic integers are denoted by O_F and O_E respectively. Let E/F be a cyclic field extension of degree n with Galois group $\mathrm{Gal}(E/F) = \langle \sigma \rangle$, where σ is the generator of the cyclic group. Let $\mathcal{A} = (E/F, \sigma, \gamma)$ be the corresponding cyclic algebra of degree n (n is also called the *index* of $\mathcal{A}$ and in practice it determines the number of transmitters), that is $\mathcal{A} = E \oplus uE \oplus u^2 E \oplus \cdots \oplus u^{n-1}E$, with $u \in \mathcal{A}$ such that $eu = u\sigma(e)$ for all $e \in E$ and $u^n = \gamma \in F^*$. An element $x = x_0 + ux_1 + \cdots + u^{n-1}x_{n-1} \in \mathcal{A}$ has the following left regular representation

as a matrix $A =$
$$
\begin{pmatrix}
x_0 & \gamma\sigma(x_{n-1}) & \gamma\sigma^2(x_{n-2}) & \cdots & \gamma\sigma^{n-1}(x_1) \\
x_1 & \sigma(x_0) & \gamma\sigma^2(x_{n-1}) & & \gamma\sigma^{n-1}(x_2) \\
x_2 & \sigma(x_1) & \sigma^2(x_0) & & \gamma\sigma^{n-1}(x_3) \\
\vdots & & & & \vdots \\
x_{n-1} & \sigma(x_{n-2}) & \sigma^2(x_{n-3}) & \cdots & \sigma^{n-1}(x_0)
\end{pmatrix}.
$$

We often identify an element x with its representation A. All algebras considered here are finite dimensional associative algebras over a field.

Definition 2.1. An algebra $\mathcal{A}$ is called *simple* if it has no nontrivial ideals. An F-algebra $\mathcal{A}$ is *central* if its center $Z(A) = \{a \in \mathcal{A} | aa' = a'a \; \forall a' \in \mathcal{A}\} = F$.

Definition 2.2. The determinant (resp. trace) of the matrix A is called the *reduced norm* (resp. *reduced trace*) of an element $a \in \mathcal{A}$ and is denoted by $nr(a)$ (resp. $tr(a)$).

The next proposition tells us when an algebra is a division algebra. Having a division algebra is crucial in order to satisfy the rank criterion.

Proposition 2.1. ([7, Theorem 11.12, p. 184]) *The algebra* $\mathcal{A} = (E/F, \sigma, \gamma)$ *of degree n is a division algebra if and only if the smallest factor $t \in \mathbb{Z}_+$ of n such that γ^t is the norm of some element in E^* is n.*

We are now ready to present some of the basic definitions and results from the theory of maximal orders. The general theory of maximal orders can be found in [8].

Let R denote a Noetherian integral domain with a quotient field F, and let $\mathcal{A}$ be a finite dimensional F-algebra. Due to practical reasons, R is usually chosen to be either $\mathbb{Z}[i]$ or $\mathbb{Z}[\omega]$ ($\omega^3 = 1$), with the respective quotient field $\mathbb{Q}(i)$ or $\mathbb{Q}(\omega)$.

Definition 2.3. An *R-order* in the F-algebra $\mathcal{A}$ is a subring Λ of $\mathcal{A}$, having the same identity element as $\mathcal{A}$, and such that Λ is a finitely generated module over R and generates $\mathcal{A}$ as a linear space over F.

As usual, an R-order in $\mathcal{A}$ is said to be *maximal*, if it is not properly contained in any other R-order in $\mathcal{A}$. If the integral closure (cf. [8]) $\overline{R}$ of R in $\mathcal{A}$ happens to be an R-order in $\mathcal{A}$, then $\overline{R}$ is automatically the unique maximal R-order in $\mathcal{A}$.

Proposition 2.2. ([8, Theorem 10.1, p. 125]) *Let Λ be an R-order in $\mathcal{A}$. For each $a \in \Lambda$ we have $nr(a) \in R$, $tr(a) \in R$. Especially, if R is the ring of integers or an imaginary quadratic number field, then $nr(a) \geq 1$, and hence the NVD property is achieved.*

Next we describe an order from where the elements are drawn in a typical CDA based MIMO space-time block code. Some optimization to this can be done e.g. with the aid of ideals as in [9] or by using a maximal order [1]. Already in [10, 11] maximal orders were successfully used for designing single receiver (MISO) space-time codes.

Definition 2.4. In any cyclic algebra where the element $\gamma \in F^*$ determining the 2-cocycle in $H^2(E/F)$ (cf. [5]) happens to be an algebraic integer, we define the following *natural order* $\Lambda_N = \mathcal{O}_E \oplus u\mathcal{O}_E \oplus \cdots \oplus u^{n-1}\mathcal{O}_E$. It is easy to show that the element γ can always be chosen (up to isomorphism of the resulting algebras) to be an algebraic integer.

Later on Theorems 2.1 and 2.3 will show us that replacing a natural order by a maximal one will result in a denser lattice with no penalty in the minimum determinant. Hence, using a maximal order is desirable.

Hereafter, F will be an algebraic number field and R a Dedekind ring with F as a field of fractions.

Proposition 2.3. *Let Γ be a subring of $\mathcal{A}$ containing R such that $F\Gamma = \mathcal{A}$, and suppose that each $a \in \Gamma$ is integral over R. Then Γ is an R-order in $\mathcal{A}$. Conversely, every R-order in $\mathcal{A}$ has these properties.*

Corollary 2.1. *Every R-order in $\mathcal{A}$ is contained in a maximal R-order in $\mathcal{A}$. There exists at least one maximal R-order in $\mathcal{A}$.*

Proposition 2.4. *Let $\mathcal{A}$ be a finite dimensional semisimple algebra over F and Λ be a $\mathbb{Z}$-order in $\mathcal{A}$. Let O_F stand for the ring of algebraic integers of F. Then $\Gamma = O_F\Lambda$ is an O_F-order containing Λ. As a consequence, a maximal $\mathbb{Z}$-order in $\mathcal{A}$ is a maximal O_F-order as well.*

Definition 2.5. Let $m = dim_F \mathcal{A}$. The *discriminant* of the R-order Λ is the ideal $d(\Lambda/R)$ in R generated by the set $\{det(tr(x_i x_j))_{i,j=1}^m \mid (x_1, ..., x_m) \in \Lambda^m\}$. It is clear that if $\Lambda \subseteq \Gamma$ then $d(\Gamma/R)|d(\Lambda/R)$. Moreover, in this case $\Lambda = \Gamma$ if and only if $d(\Gamma/R) = d(\Lambda/R)$. As maximal orders all share the same discriminant, we can conclude that within a fixed algebra, maximal orders have the smallest discriminant. We will call this the *discriminant $d_{\mathcal{A}}$ of the algebra $\mathcal{A}$*.

If R is a principal ideal domain, then $d(\Lambda/R) = det(tr(c_i c_j))R$, where $\{c_1, ..., c_m\}$ is a basis of Λ over R. In this case, $d(\Lambda/R)$ can be identified with an element $\beta \in R$, which generates the principal ideal $d(\Lambda/R)$. Clearly, β is unique up to multiplication by a unit of R.

In [1], we proved the following convenient connection between the discriminant of an order and the measure of the fundamental parallelotope of the corresponding lattice. The result is somewhat unsurprising, as the definition of the discriminant closely resembles that of the Gram matrix.

Theorem 2.1. *Assume that F is an imaginary quadratic number field and that 1 and θ form a $\mathbb{Z}$-basis of its ring of integers R. Assume further that the order Λ is a free R-module (an assumption automatically satisfied, when R is a principal ideal domain). Then the measure of the fundamental parallelotope equals $m(\Lambda) = |\Im\theta|^{n^2} |d(\Lambda/R)|$.*

In the respective cases $F = \mathbf{Q}(i)$ and $F = \mathbf{Q}(\sqrt{-3})$ we have $\theta = i$ and $\theta = (-1 + \sqrt{-3})/2$. Whence, the following two corollaries are immediate.

Corollary 2.2. *Let $F = \mathbf{Q}(i), R = \mathbb{Z}[i]$, and assume that $\Lambda \subset (E/F, \sigma, \gamma)$ is an R-order. Then the measure of the fundamental parallelotope equals $m(\Lambda) = |d(\Lambda/\mathbb{Z}[i])|$.*

Corollary 2.3. *Let $\omega = (-1 + \sqrt{-3})/2$, $F = \mathbf{Q}(\omega)$, $R = \mathbb{Z}[\omega]$, and assume that $\Lambda \subset (E/F, \sigma, \gamma)$ is an R-order. Then the measure of the fundamental parallelotope equals $m(\Lambda) = (\sqrt{3}/2)^{n^2} |d(\Lambda/\mathbb{Z}[\omega])|$.*

3. The discriminant bound

In this section, we review some useful results from [1]. Again let F be an algebraic number field that is finite dimensional over $\mathbf{Q}$, $\mathcal{O}_F$ its ring of integers, P a prime ideal of $\mathcal{O}_F$ and $\hat{F}_P$ the completion. In what follows we discuss the size of ideals of $\mathcal{O}_F$. By this we mean that ideals are ordered by the absolute values of their norms to $\mathbf{Q}$, so e.g. in the case $\mathcal{O}_F = \mathbb{Z}[i]$ we say that the prime ideal generated by $2 + i$ is smaller than the prime ideal generated by 3 as they have norms 5 and 9, respectively.

The following relatively deep result from class field theory is the key for deriving the discriminant bound. Assume that the field F is totally complex. Then we have the *fundamental exact sequence of Brauer groups* (see e.g. [8] or [6]) $0 \longrightarrow \mathrm{Br}(F) \longrightarrow \oplus \mathrm{Br}(\hat{F}_P) \longrightarrow \mathbf{Q}/\mathbf{Z} \longrightarrow 0$.

Here, the first nontrivial map is obtained by mapping the similarity class of a central division F-algebra $\mathcal{D}$ to a vector consisting of the similarity classes of all the simple algebras $\mathcal{D}_P$ obtained from $\mathcal{D}$ by extending the scalars from F to $\hat{F}_P$, where P ranges over all the prime ideals of $\mathcal{O}_F$. Observe that $\mathcal{D}_P$ is not necessarily a division algebra, but by Wedderburn's theorem [5, p. 203] it can be written in the form $\mathcal{D}_P = \mathcal{M}_{\kappa_P}(\mathcal{A}_P)$, where $\mathcal{A}_P$ is a division algebra with a center $\hat{F}_P$, and κ_P is a natural number called the *local capacity*. The second nontrivial map of the fundamental exact sequence is then simply the sum of the Hasse invariants of the division algebras $\mathcal{A}_P$ representing elements of the Brauer groups $\mathrm{Br}(\hat{F}_P)$.

This exact sequence tacitly contains the piece of information that for all but finitely many primes P the resulting algebra $\mathcal{D}_P$ is actually in the trivial similarity class of $\hat{F}_P$-algebras. In other words $\mathcal{D}_P$ is isomorphic to a matrix algebra over $\hat{F}_P$. More importantly, the sequence tells us that the sum of the nontrivial Hasse invariants of any central division algebras must be an integer. Furthermore, this is the only constraint for the Hasse invariants, i.e. any combination of Hasse invariants (a/m_P) such that only finitely many of them are non-zero, and that they sum up to an integer, is realized as a collection of the Hasse invariants of some central division algebra $\mathcal{D}$ over F.

Let us now suppose that with a given number field F we would like to produce a division algebra $\mathcal{A}$ of a given index n, having F as its center and the smallest possible discriminant. We proceed to show that while we cannot give an explicit description of the algebra $\mathcal{A}$ in all the cases, we can derive an explicit formula for its discriminant. For the proof, see [1].

Theorem 3.1. *Assume that the field F is totally complex and that $P_1, \ldots, P_n$ are some prime ideals of $\mathcal{O}_F$. Assume further that a sequence of rational numbers $a_1/m_{P_1}, \ldots, a_n/m_{P_n}$ satisfies $\sum_{i=1}^n \frac{a_i}{m_{P_i}} \equiv 0 \pmod{1}$, $1 \le a_i \le m_{P_i}$, and $(a_i, m_{P_i}) = 1$. Then there exists a central division F-algebra $\mathcal{A}$ that has local indices m_{P_i} and the least common multiple (LCM) of the numbers $\{m_{P_i}\}$ as an index. Further, if Λ is a maximal $\mathcal{O}_F$-order in $\mathcal{A}$, then the discriminant of Λ is $d(\Lambda/\mathcal{O}_F) = \prod_{i=1}^n P_i^{(m_{P_i}-1)\frac{[\mathcal{A}:F]}{m_{P_i}}}$.*

At this point, it is clear that the discriminant $d_{\mathcal{A}}$ of a division algebra only depends on its local indices m_{P_i}.

Now we have an optimization problem to solve. Given the center F and an integer n we should decide how to choose the local indices and the Hasse invariants so that the LCM of the local indices is n, the sum of the Hasse invariants is an integer, and that the resulting discriminant is as small as possible. We immediately observe that at least two of the Hasse invariants must be non-integral.

Observe that the exponent $d(P)$ of the prime ideal P in the discriminant formula $d(P) = (m_P - 1)\frac{[\mathcal{A}:F]}{m_P} = n^2\left(1 - \frac{1}{m_P}\right)$. As for the nontrivial Hasse invariants $n \ge m_P \ge 2$, we see that $n^2/2 \le d(P) \le n(n-1)$. Therefore the nontrivial exponents are roughly of the same size. E.g. when $n = 6$, $d(P)$ will be either 18, 24 or 30 according to whether m_P is 2, 3 or 6. Not surprisingly, it turns out that the optimal choice is to have only two non-zero Hasse invariants and to associate these with the two smallest prime ideals of

$\mathcal{O}_F$. The following discriminant bound [12] thanks to Dr. Roope Vehkalahti now gives us a great insight to the minimization of the discriminant, i.e. maximization of the minimum determinant.

Theorem 3.2 ([1] Discriminant bound). *Assume that F is a totally complex number field, and that P_1 and P_2 are the two smallest prime ideals in $\mathcal{O}_F$. Then the smallest possible discriminant of all central division algebras over F of index n is $(P_1 P_2)^{n(n-1)}$.*

We remark that in the most interesting (for MIMO) cases $n = 2$ and $n = 3$, the proof of Theorem 3.2 is more or less an immediate corollary of Theorem 3.1. We also remark that the division algebra achieving our bound is by no means unique. E.g. any pair of Hasse invariants $a/n, (n - a)/n$, where $0 < a < n$, and $(a, n) = 1$, leads to a division algebra with the same discriminant.

The smallest primes of the ring $\mathbf{Z}[i]$ are $1+i$ and $2\pm i$. They have norms 2 and 5 respectively. The smallest primes of the ring $\mathbf{Z}[\omega]$ are $\sqrt{-3}$ and 2 with respective norms 3 and 4. Together with Corollaries 2.2 and 2.3 we have arrived at the following bounds.

Corollary 3.1 (Discriminant bound). *Let Λ be an order of a central division algebra of index n over the field $\mathbf{Q}(i)$. Then the measure of a fundamental parallelotope of the corresponding lattice $m(\Lambda) \geq 10^{n(n-1)/2}$.*

Corollary 3.2 (Discriminant bound). *Let Λ be an order of a central division algebra of index n over the field $\mathbf{Q}(\omega)$, $\omega = (-1+\sqrt{-3})/2$. Then the measure of a fundamental parallelotope of the corresponding lattice $m(\Lambda) \geq (\sqrt{3}/2)^{n^2} 12^{n(n-1)/2}$.*

Example 2. In the so-called Golden division algebra [9], i.e. the cyclic algebra $\mathcal{GA} = (E/F, \sigma, \gamma)$ obtained from the data $E = \mathbf{Q}(i, \sqrt{5})$, $F = \mathbb{Q}(i)$, $\gamma = i$, $n = 2$, $\sigma(\sqrt{5}) = -\sqrt{5}$, the natural order Λ_N is already maximal [11]. Therefore, within the algebra $\mathcal{GA}$, no further optimization can be done when dealing with orders. Instead, the authors of [9] do the optimization by using an ideal $\mathcal{I}$ of norm 5 generated by the $\mathbb{Z}[i]$-basis $\{\alpha, \alpha\theta\}$, where $\theta = (1 + \sqrt{5})/2$ and $\alpha = 1+i-i\theta$. The ideal does not affect the density, but improves the shape of the lattice as it gives an orthogonal basis. The code matrices of the Golden code are then of the form $\begin{pmatrix} \alpha(a + b\theta) & \gamma\sigma(\alpha(a + b\theta)) \\ \alpha(c + d\theta) & \sigma(\alpha(c + d\theta)) \end{pmatrix}$, where $a, b, c, d \in \mathbb{Z}[i]$. The Golden code is a special case of the so-called perfect codes, see [9]. The $\mathbb{Z}[i]$ discriminant of the Golden code is 25, whereas the minimal possible discriminant (cf. Corollary 3.1) would be 10. Thus, the Golden code is not optimal with respect to density.

Example 3. On the other hand, in the Golden+ division algebra [1], i.e. the cyclic algebra $\mathcal{GA}+ = (E/F, \sigma, \gamma)$ obtained from the data $E = \mathbf{Q}(s = \sqrt{2+i})$, $F = \mathbb{Q}(i)$, $\gamma = i$, $n = 2$, $\sigma(s) = -s$, the natural order Λ_N is not maximal. A maximal order can be produced by hand, following a construction algorithm due to Ivanyos and Rnyai. The algorithm is implemented in the MAGMA computer algebra software, so alternatively one can use the free online calculator to more easily produce the basis for a maximal order. See [1] for a detailed description of the Golden+ code, which is built from a maximal order of $\mathcal{GA}+$. The Golden+ code does have the minimal discriminant $= 10$. Thanks to its higher (optimal) density, the Golden+ code performs better than the celebrated Golden code.

References

1. R. Vehkalahti, C. Hollanti, J. Lahtonen and K. Ranto, "On the densest MIMO lattices from cyclic division algebras", *IEEE Trans. on Inform. Theory* (in press), 2008. *http://arxiv.org/abs/cs.IT/0703052*.
2. J.-C. Guey, M. P. Fitz, M. R. Bell and W.-Y. Kuo, Signal design for transmitter diversity wireless communication systems over rayleigh fading channels, in *Proc. IEEE VTC'96*, 1996.
3. V. Tarokh, N. Seshadri and A. R. Calderbank, "Space-time codes for high data rate wireless communication: Performance criterion and code construction", *IEEE Trans. Inf. Theory* **44**, 744(Mar. 1998).
4. B. A. Sethuraman, B. S. Rajan and V. Shashidhar, "Full-diversity, high-rate space-time block codes from division algebras", *IEEE Trans. Inf. Theory* **49**, 2596(Oct. 2003).
5. N. Jacobson, *Basic Algebra II* (W. H. Freeman and Company, San Francisco, 1980).
6. J. S. Milne, Class field theory Lecture notes for a course given at the University of Michigan, Ann Arbor, *http://www.jmilne.org/math/coursenotes/*.
7. A. A. Albert, *Structure of Algebras* (American Mathematical Society, New York, 1939).
8. I. Reiner, *Maximal Orders* (Academic Press, New York, 1975).
9. F. Oggier, G. Rekaya, J.-C. Belfiore and E. Viterbo, "Perfect space-time block codes", *IEEE Trans. Inf. Theory* **52**, 3885(Sept. 2006).
10. C. Hollanti, J. Lahtonen and H.-F. Lu, "Maximal Orders in the Design of Dense Space-Time Lattice Codes", *IEEE Trans. Inf. Theory* **54**, 4493 (2008).
11. C. Hollanti and J. Lahtonen, A new tool: Constructing STBCs from maximal orders in central simple algebras, in *Proc. 2006 IEEE Inform. Theory Workshop*, (Punta del Este, Uruguay, 2006).
12. R. Vehkalahti, Class field theoretic methods in the design of lattice signal constellations, PhD thesis2008. *TUCS Dissertations Series*, no. 100, *https://oa.doria.fi/handle/10024/36604*.

PART C

Women in Mathematics

AND WHAT BECAME OF THE WOMEN?

C. SERIES

Mathematics Department, University of Warwick
Coventry, CV4 7AL, UK
C.M.Series@warwick.ac.uk
www.warwick.ac.uk/staff/C.M.Series

This article is about the lives of three women, Philippa Fawcett, Charlotte Scott and Grace Young, who studied mathematics in Cambridge University in the period 1875–1895. It was first published in *Mathematical Spectrum* in 1997.* It was written in response to an earlier article in the same magazine entitled *What became of the Senior Wranglers?*, in which only one woman was mentioned. The 'Senior Wrangler' was the person achieving top marks in the final mathematics examinations, known as 'The Mathematics Tripos', at Cambridge.

Keywords: Women mathematicians; Cambridge 1875–1895; Mathematics Tripos.

Hail the triumph of the corset
Hail the fair Philippa Fawcett.
Victress in the fray.
Crown her queen of hydrostatics
And the other Mathematics
Wreathe her brow in bay.

".... the University would think the examination of
young ladies a matter altogether beyond its sphere of duty."

Oxford Local Examination Delegacy, 1863.

Introduction

The September 1996 issue of Mathematical Spectrum contained an article "What became of the Senior Wranglers?" by D. O. Forfar. In that article, only one woman, Philippa Fawcett, is mentioned. Since during the period covered by the article, 1753–1909, women were not allowed to take degrees at Cambridge, and since the women's colleges at Girton and Newnham were only established in 1869 and 1871 respectively, this omission is scarcely surprising. When one considers the lamentable state of women's education up till the late years of the nineteenth century, and the enormous public prejudice which existed against women studying science or mathematics, it is really much more surprising that, towards the end of that period, there were several women who, morally at least, did attain Wranglerhood. At the time their achievements were hailed as turning points in the struggle for women's education. Today their stories can still inspire.

At Cambridge

In 1890, Philippa Fawcett scored the highest mark of all candidates in Part 1 of the Mathematical Tripos. She was placed "above the Senior Wrangler". Her triumph, in mathematics, that last bastion of superiority of the male mind, was spectacular. There was lengthy discussion and comment in national papers in England and abroad. It would have been hard to think of a more effective or timely challenge to popular prejudice. Women's powers of reasoning could no longer be said to be inferior to men's.

Shortly before Philippa's birth in 1868, the pressure for some provision for education of women at Oxford and Cambridge was growing. Henry Sidgwick, Professor of Moral Philosophy at Cambridge, and one of the driving forces in this movement, started a series of meetings, often held in the Fawcetts' drawing room, to make plans. Her mother's diary records: "Philippa was aged about two at this time, old enough to be brought in at the tea-drinking stage at the end of the proceedings and to toddle about in her white frock and blue sash amongst the guests."

Anne Jemima Clough, born in 1820 and brought into contact with the campaign for women's education by her brother the poet Arthur Clough, had been running a scheme of lectures for senior girls which by 1869 had spread to some 25 centres in the north of England. At a momentous meeting held in the Fawcetts' house in December 1869, it was decided to set up a similar scheme of lectures in Cambridge, and in the spring of 1870 a series of lectures was attended by 70–80 women. Provision needed to be made for a "hall or lodging" for women wishing to attend from outside Cambridge. Proceeding with the utmost discretion, the following summer Sidgwick rented and furnished a house at his own expense and invited Miss Clough to take charge with the first five students. This was the beginning of Newnham College.

Emily Davies, born in 1830, had been a tireless campaigner for women's education for many years. It was she who inspired Philippa's aunt Elizabeth Garrett Anderson to study medicine seriously and who stood by her side through ten years of struggle which paved the way for the admission of women into the medical profession. Described by a contemporary as "a rather dim little person with mouse coloured hair and conventional manners", she was single-minded and ruthless, inflexible in her view that women could only challenge men's intellectual dominance if they matched them at their own tests. It was Emily Davies who persuaded the Cambridge Local Examination Syndicate to agree to a trial examination for girls. Despite having only six weeks to prepare, the performance of the 83 girls was found comparable to that of the boys in all subjects except arithmetic. (The quality of teaching improved so much as result of this embarrassing discovery that within three years no inferiority could be detected!)

As a result of Emily Davies' efforts, Girton College for women opened in 1869 at Hitchin, midway between Cambridge and London, also with five students. One of the first scholars, Miss Woodhead, daughter of a Quaker grocer, studied mathematics. She was tutored by Mr. Stuart, later Cambridge Professor of Mechanics, and Mr. J. L. Moulton, Senior Wrangler and Smith's prizeman, later Lord Justice Moulton, who, it is recorded, poured "amazing illuminations on elementary mathematics".

Notwithstanding Miss Davies' efforts, the University Council refused to admit Girton students to University examinations, although they "carefully abstained from expressing any disapproval of the Examiners' examining the students in their private capacity and in a clandestine way." Thus in 1872 the first three candidates for the Tripos, including Miss Woodhead, were chaperoned into Cambridge by Miss Davies and took the examination in the sitting room of the University Arms. Despite the papers arriving an hour late (the runner was given the wrong address), the three passed with flying colours. When the news reached Hitchin, elated young women climbed onto the roof and rang the alarm bell so loudly that fire engines were got out.

In 1873 Girton College moved to its present location on the edge of Cambridge, where already more than half the professors admitted women to their lectures. Special lectures were still given under the Sidgwick arrangement, and other lecturers cycled out to Girton to coach. In a very short time, Newnham and Girton students began to challenge old prejudices and in particular to challenge men in the examinations. Selected girls were allowed to take the University examinations, but this was only by special permission, not by right, nor were their names to be included in the lists of results. Nevertheless, between 1874 and 1881, 21 students had entered for a Tripos examination and all had succeeded, four having been placed in the first class.

In 1876, the 18 year old Charlotte Angas Scott was awarded a scholarship to Girton to study mathematics "on the basis of home tutoring." Charlotte had had no formal schooling but her father, Caleb Scott, a dynamic man and president of a nonconformist college near Manchester, had doubtless encouraged his daughter's studies. The entering class contained 11 girls.

Charlotte took the Mathematics Tripos in January 1880. Campaigning by Emily Davies had gained the girls permission to sit the same examinations as the men (in different rooms, of course). Their results would be read out after the men's, but in mathematics each candidate would be assigned a place relative to the ordered list of male candidates. The news leaked out that Charlotte had been placed eigth. Women were not allowed at the ceremony at which the results were read out, but when it came to the eighth on the list the undergraduates called out "Scott of Girton, Scott of Girton" and there was such an uproar that the poor man's name could not be heard.

Charlotte Scott was the first woman Wrangler. Her achievement, and "in a man's subject" at that, made a deep public impression. Recalling the event at her retirement celebrations 45 years later Professor Harkness, who at the time had been a schoolboy in Cambridge, said that he believed her achievement marked the turning point in England from "the theoretical feminism of Mill and others to the practical educational and political advances of the present time." So strongly was public opinion aroused that a petition with over 10,000 signatures to grant women the right to sit examinations and be admitted to degrees was presented to the Cambridge authorities. Arthur Cayley, a leading Cambridge mathematician, was one of the main supporters and Charlotte's triumph cited as one of the main grounds. After a year of public pressure, the University voted in 1881 to grant women the right to be examined and to have their names on the official class lists, though in a separate table from the men. The other request in the petition fared worse: women were not eligible for Cambridge degrees until 1948.

Despite Charlotte's achievement, it was still widely believed that mathematics was peculiarly incompatible with female thought processes. In fact 34 out of the 40 girls who entered the first preliminary trial for the Cambridge Local examination in arithmetic failed. As Henry Fawcett reputedly remarked: "he did not imagine if the Universities were opened to women they would produce any Senior Wranglers." He had not reckoned with his own daughter.

Philippa Fawcett came of a distinguished family. Her father Henry rose to be Postmaster General under Gladstone and was the man responsible for introducing the parcel post. Her mother Millicent, later Dame Millicent, was one of the leaders of the non-violent campaign for women's votes. Philippa herself was, in the words of one her Newnham contemporaries, "modest and retiring almost to a fault". She lived a very regular and quiet life and was coached by Mr E. W. Hobson of Christ's, a Senior Wrangler himself and judged to be the second best coach. She also played hockey. Philippa did outstandingly well in the exams she sat in the second year, with 75 more marks than the top Trinity man. Everyone anticipated a brilliant result in the Tripos.

The scene in the Senate when the results were to be announced is recorded in a letter written by Philippa's second cousin Marion: "...the gallery was crowded with girls and a few men...The floor was thronged by undergraduates... All the men's names were read first, the Senior Wrangler was much cheered... At last the man who had been reading shouted 'Women'. The undergraduates yelled 'Ladies' and for some moments there was a great uproar. A fearfully agitating moment for Philippa it must have been; the examiner could not attempt to read the names until there was a lull. Again and again he raised his cap, but he would not say 'ladies' instead of 'women' and quite right I think... At last he read Philippa's name, and announced she was 'above the Senior Wrangler'. There was great and prolonged cheering; many of the men turned towards Philippa, who was sitting in the gallery with Miss Clough, and raised their hats. When the examiner went on with the other names there were cries of 'Read Miss Fawcett's name again' but no attention was paid to this. I don't think any other women's names were heard, for the men were making such a tremendous noise..."

On her arrival back at College, Philippa was greeted by a crowd of fellow students and carried into Hall. Flowers, letters and telegrams poured in throughout the day. That evening there was an impromptu college feast and she was carried three times round a bonfire on the hockey pitch. The triumphal lay, whose first verse heads this article, was composed in her honour. The story made the lead in the Telegraph the next day: "Once again has woman demonstrated her superiority in the face of an incredulous and somewhat unsympathetic world... And now the last trench has been carried by Amazonian assault, and the whole citadel of learning lies open and defenceless before the victorious students of Newnham and Girton. There is no longer any field of learning in which the lady student does not excel."

The last (moral) Wrangler of this period, Grace Chisholm Young, was born in 1868. Her father, a distiguished civil servant but already almost sixty when she was born, retired when Grace was only seven, and took an active role in supervising her education at home. Grace won a scholarship to Girton 1889. She became a Wrangler in Part 1 of the Tripos in 1892. Immediately afterwards she and a friend Isabel Maddison went to Oxford and sat for the final honours school in Mathematics, according to Dame Mary Cartwright (see postscript), "just to show". Grace obtained the highest marks of all students that year and became the first person of either sex to obtain a First in any subject at both Oxford and Cambridge. Grace and Isabel were the first women to take finals in Mathematics at Oxford, and it seems that no woman did so again until Dame Mary herself in 1923.

What did these three women do afterwards?

Charlotte Scott began lecturing at Girton and began work on a doctorate under Cayley. Since Cambridge did not grant advanced degrees to women, she took a B.Sc. from the University of London by external examination in 1882, and a Ph. D. by the same route in 1885. By a great stroke of good fortune, she was almost immediately offered the job of Head of the Mathematics department at the newly founded American women's college Bryn Mawr, where she remained until her retirement in 1925. Scott came to be widely recognised as a mathematician. A first edition of *American Men of Science* shows her name starred. Her text on analytic geometry was reprinted after thirty years. She never married, but supported and encouraged generations of women mathematicians, many of whom went on to teach all over the United States. She played a leading role in American mathematical life and was widely respected as a scholar and teacher, a wise and gifted administrator, and a rock of integrity. On her retirement she returned to England and is buried in St. Giles Churchyard, Cambridge.

In 1891, Philippa Fawcett, together with the Senior Wrangler, G.T. Bennett, was placed in the top division of the first class of Part II. Philippa was awarded a scholarship at Newnham which gave her a further year of study. During this year she made her only contribution to research, a long paper on the motion of helical bodies in liquid. For the next 14 years she was a Newnham College lecturer. Then, following a trip to South Africa, she was invited to take up a post as a lecturer in a normal school in Johannesburg where she trained mathematics teachers. In 1905 she returned to England as principal assistant to the Director of Education in the newly formed London County Council. (She was, remarkably, offered this job without interview and at the same salary as a man.) She continued in this post till her retirement in 1934. She died in 1948, two months after her eightieth birthday and one month after Cambridge women were finally granted degrees.

After Oxford finals, Grace Chisholm returned to Cambridge and completed Part II of the Tripos. There was no possibility of a woman getting a Fellowship at Cambridge. However at just that time, as part of an experiment, a small group of women were to be recruited to study at Göttingen under Felix Klein, one of the leading mathematicians of the day. So as to be on the safe side and so as to establish no unwelcome precedents, the women were to be foreigners, and, just to be sure, their subject would be mathematics. As Klein explained later : "Mathematics had here rendered a pioneering service to the other disciplines. With it matters are, indeed, most straightforward. In mathematics, deception as to whether real understanding is present or not is least possible."

Thus Grace Chisholm became one of three women admitted to Göttingen in 1893. They had to behave very discretely: "We are to go to Prof. Klein's private office before the regular time for changing classes so as not to meet the students in the halls and from there we are to go into the class."

All went smoothly and Grace became the first woman in East Prussia to gain a Ph.D., which she did *magna cum laude* in 1895. She returned to England and married one of her former tutors, W. H. Young. Following a visit of Klein in 1897, they decided to "throw up filthy lucre, go abroad, and devote ourselves to research." Until their marriage, Young seems not to have done any research. Together, however, they began to publish many papers, and it seems probable that Grace's contributions, even to those written under her husband's name alone, were considerable. Their work was strongly influenced by the new ideas with which Grace had come into contact in Germany, and had in turn had a strong influence in England, helping to establish the new standard of rigour which was the foundation of Cambridge's reputation as a world centre of pure mathematics.

Grace had very many interests outside mathematics and even found time to achieve her ambition of studying medicine. She brought up six children, two of whom themselves became mathematicians. Her husband died in 1942 two years before Grace herself.

Postscript

In the second half of the 20th century, education for women has become the norm. There is now a good sprinkling of women mathematicians in university posts although not so many in the senior ranks. Only two have achieved the distinction of being an Fellow of the Royal Society: Dame Lucy Mary Cartwright (1900–)[a], Mistress of Girton 1949–68, whose work straddles the 20th century and was foundational for the modern theory of chaos and, much more recently, Dusa McDuff (1945–), elected an FRS in 1994 in recognition of her work on symplectic geometry.[b]

Sources

The main sources for this article are listed below.

References

1. M. L. Cartwright, *Grace Chisholm Young, Obituary, J. London Math. Soc.* **19**, 185–192 (1944).
2. M. L. Cartwright, *Non-linear vibrations: A Chapter in Mathematical History, Mathematical Gazette* **316**, 81–89 (1952).
3. J. Green and J. LaDuke, *Women in the American Mathematical Community, Mathematical Intelligencer* **9(1)**, 11–23 (1987).
4. Patricia C. Kenschaft, *Charlotte Angas Scott 1858–1931, AWM Newsletter* **7(6)**, (1977) and **8(1)**, (1978).
5. Patricia C. Kenschaft, *Charlotte Angas Scott 1858-1931*, in *Women and Mathematics* L. Grinstein and P. J. Campbell eds, (Greenwood Press 1987).
6. Rota McWilliams-Tullberg, *Women at Cambridge*, (Gollancz 1975).
7. P. Rothman, *Grace Chisholm Young and the division of laurels, Notes Rec. Roy. Soc. London* **50(1)**, 89–100 (1996).
8. Emily James Putnam, *Celebration in Honour of Professor Scott, Bryn Mawr Alumni Bulletin* **II(5)**, (1922).

---

[a]Dame Mary Cartwright died in April 1998.

[b]Since this article was written, three further women mathematicians have been elected Fellows of the Royal Society: Frances Kirwan (2001), Mary Rees (2002), and Ulrike Tillmann (2008).

9. Steven Siklos, *Philippa Fawcett and the Mathematical Tripos*, (Newnham College, Cambridge 1990).

10. Barbara Stephen, *Emily Davies and Girton College*, (Constable and Co, London 1927).

11. Barbara Stephen, *Girton College 1869-1932*, (Cambridge University Press, 1933.)

THREE GREAT GIRTON MATHEMATICIANS

RUTH M. WILLIAMS

Department of Applied Mathematics and Theoretical Physics
Centre for Mathematical Sciences, Wilberforce Road
Cambridge CB3 0WA, England
and
Girton College, Cambridge CB3 0JG, England
rmw7@damtp.cam.ac.uk

Girton was the first Cambridge college to be established for women, and for just over the first hundred years of its existence, it remained a women's college. Throughout its history, it has had a number of distinguished women mathematicians, and this article looks briefly at three whose lives spanned almost the whole of the twentieth century, and whose interests were in three different areas of mathematics.

Keywords: Women mathematicians; Girton College.

Dame Mary Cartwright, F.R.S.

Born on December 17, 1900 in Aynho, Northamptonshire, where her father was curate and later rector, Mary Lucy Cartwright was at first educated by governesses and then sent away to a variety of schools. In 1919, she went up to St. Hugh's College, Oxford, to read mathematics. She felt ill-prepared for this because of gaps in her schooling and after two years when she obtained a Second in "Mods", she considered changing to history, but decided against it because it seemed to entail longer hours of work. Instrumental in her mathematical career was a chance meeting at a party on a barge on the Thames in her third year, when V. C. Morton (later to become Professor of Mathematics at Aberystwyth) suggested she start attending the evening classes of the distinguished mathematician G. H. Hardy. Quite outside the normal syllabus, these classes met on Mondays after dinner, beginning with a talk and continuing with mathematical discussions until late. Stimulated by the classes, Mary went on to obtain a First in her Finals in 1923.

On graduation, not wishing to impose further on her father for money to support her as a research student, Mary went into school teaching. However, after a few years she could no longer resist the attractions of mathematical research and returned to Oxford early in 1928 to work as a student of Hardy, who was soon very impressed with her work on complex analysis. Again there was an influential evening class, this time on Fridays, though all the research students went to the Monday class as well. She first met J. E. Littlewood (who with Hardy dominated British mathematics for much of the twentieth century) as the external examiner for her D.Phil. He records that the first question asked by the other examiner was so silly and unreal as to make her blush but "I was able to get in a wink, and I think it restored her nerve".

In 1930, after obtaining her D.Phil., Mary moved to Cambridge to a Yarrow Research Fellowship at Girton College, continuing to work on the theory of functions. The important results that she obtained, published in 1935 in the Mathematische Annalen, prompted Hardy (who had moved to Cambridge in 1931) and Littlewood to recommend her for an Assistant Lectureship in the Faculty of Mathematics. She became Lecturer in 1935 and Reader in the Theory of Functions in 1959. In the meantime, at Girton, she was a College Lecturer and Director of Studies in Mathematics until 1949. Her undergraduate pupils were at first apt to find her rather fragile and timid, an impression which was rapidly dispelled on closer acquaintance - and perhaps also on seeing her figure-skating on the college pond in severe winters. Her research students, initially in awe of her intellectual reputation, soon thawed in the warmth of her friendship and understanding. She, who in her first year in Oxford had apparently found difficulty in her own mathematical studies, sympathised with her students' weaknesses.

Mary's research career took an interesting turn in 1939 with a request from the Department of Scientific and Industrial Research to pure mathematicians for help with problems connected with defence, in particular solving "certain very objectionable-looking differential equations occuring in connection with radar". Mary who was in the habit of showing Littlewood anything which she thought would interest him, described to him the relevant work of van der Pol, and soon they were translating problems involving radio waves and oscillations into problems in dynamics. This work was a major part of what became a long, highly fruitful and harmonious collaboration, conducted mainly by letter. Littlewood once described her as "the only woman in my life to whom I have written twice in one day!". Their ground-breaking results on the periodicity and stability of solutions of non-linear differential equations form the basis for the modern theory of dynamical systems and chaos.

During this particularly creative period of work with Littlewood, Mary carried a heavy load of teaching. She was an excellent supervisor of research students, taking great care in reading and criticising their work, both in content and in form, and giving due encouragement.

In 1947, after the Statutes of the Royal Society were amended to make it explicit that women were eligible as Fellows, Mary was the fourth to be elected (and the first woman mathematician). Soon afterwards, at the height of her mathematical research career, the focus of her life changed dramatically when in October 1949 she became Mistress of Girton. She provided quiet, unassuming, clear-headed and shrewd leadership in a time of many changes; Girton, as a women's college at that time, had only just become fully incorporated into the University. Inevitably she was no longer able to devote so much time to mathematics, but she did not seem to regret this, believing the subject to be predominantly a younger person's game. Even so, honours continued to be heaped upon her, including honorary doctorates from many universities. The Royal Society bestowed its Sylvester Medal in 1964, and four years later, the London Mathematical Society, of which she had been President from 1961 to 1963, awarded her its De Morgan Medal. She was made a Dame of the British Empire in 1969.

Throughout her high-powered academic career, Mary took a great interest in mathematical teaching in schools and was President of the Mathematical Association in 1951-2. Although her published work is nearly all severely technical, she could also appeal to a wider public as she showed in her James Bryce Memorial Lecture, "The Mathematical Mind", given at Somerville College in 1955.

When she retired from being Mistress of Girton in 1968, she held visiting professorships at universities in the United States and Europe before returning to Cambridge where she was one of the editors of "The Collected Papers of G. H. Hardy". The apparent shyness and austerity of her days as Mistress disappeared in her retirement; she became much more gregarious and was often to be found eating lunch with the younger Girton Fellows. She had a wry sense of humour and the television documentary "Our Brilliant Careers", made when she was in her mid-nineties, captured the sharp sparkle of her wit. She was no narrow specialist, being exceptionally well-informed on a wide variety of subjects, among them painting and music. She had a great sense of fun and a capacity for sympathy which perhaps only her friends knew. In human affairs, as in mathematics, she had a gift for going to the heart of a matter and for seeing what was really important. In her later years, she had a number of falls resulting in broken bones, but took her long spells in hospital very philosophically. On one occasion, after she fell while visiting her brother in Wales, she had to be transported by helicopter to hospital in Cambridge and she did the map-reading for the pilot during the journey. In her nineties, she moved to a local nursing home. Frail in body but not in mind, she fascinated her visitors with reminiscences of her travels and the many distinguished mathematicians whom she had known. Many of the more than ninety mathematical papers which she published are of seminal importance. She died on April 3, 1998.

Bertha Swirles, Lady Jeffreys

Bertha Swirles was born in Northampton on 22 May 1903. The earliest surviving photograph shows her sitting up in her pram, wearing glasses, at the age of six months. Her father, a leather salesman, died when she was still a toddler, but, although she had no siblings, her childhood was far from solitary. She grew up in a large extended family, dominated by school teachers (her mother and seven of her nine aunts) so she was involved with the world of education from her earliest days. In fact, she claimed that she learned as much from her self-educated grandfather and from another aunt as she did from the professionally trained teachers. There was no scientific background in her family, but her mother subscribed to Scientific American for her; that, coupled with the fact that she was brought up mainly by women, meant that it never occurred to her that there was anything strange for a woman to want to become a mathematician or a physicist.

From the primary school where her mother taught and her aunt was headmistress, Bertha went to the newly established Northampton School for Girls, where she was taught by three Cambridge women mathematicians. Mathematics soon became her main interest, although she also liked learning languages and was apparently furious when told by a boy who was learning Russian that it was not a suitable language for a girl.

In 1921, Bertha went as a major scholar to Girton to read mathematics, obtaining a first in both parts of the Tripos and graduating in 1924. Her interest in physics had been encouraged by a woman research student of Appleton, Mary Taylor, who was to become an important role model for her, and she spent the next academic year studying Part II Physics, attending lectures by J. J. Thompson and Rutherford, among others. Having found financial support for research through a Yarrow Studentship from Girton and a DSIR grant, she became a research student of Fowler in 1925, but in fact her first research problem came from Douglas Hartree. When she was shown into a room with two pianos, she thought to herself "This is the place for me!" and it was indeed the beginning of a life-long friendship with the Hartrees. Soon she became involved in research in the new and very exciting field of quantum theory. The physics societies which met in the evenings were not open to women, but she did read a paper in 1927 at a joint meeting of St John's Mathematical Society, the Adams Society, and the Girton Mathematical Society (which, she was quick to point out, was much older!).

In 1927, Girton elected Bertha to a Hertha Ayrton Research Fellowship, and, encouraged by Fowler and Mary Taylor, she spent the winter semester of 1927-8 in Göttingen, which was then the epicentre of research in quantum theory. It was a tremendously stimulating time to be there; she studied under Born and Heisenberg, and interacted with many of the other leading continental workers in the field. She returned to Cambridge for the Easter Term 1928 to finish her thesis, "On some applications of the theory of perturbations in quantum mechanics", and by the time she was awarded her Ph.D. in 1929, she was already an Assistant Lecturer in Manchester. She held similar posts briefly at Bristol and Imperial College before returning to Manchester in 1933 as a Lecturer in Applied Mathematics, and perhaps her most important research was done at this time. It was prompted by a remark by Hartree during an encounter on Euston Station in 1934, which led to a series of papers on multi-electron atoms, with increasingly good approximations, culminating in a classic paper, published in 1939, with Douglas Hartree and his father, who did the numerical work on a Brunsviga hand calculating machine. Bertha returned to Cambridge in 1938, to an Official Fellowship and Lectureship in Mathematics at Girton, where she remained a Fellow for the rest of her life.

Although much of Bertha's energy went into teaching, she continued with her research in quantum theory. However her most widely-known and influential publication was the book "Methods of Mathematical Physics", written with the distinguished mathematician and geophysicist, Harold Jeffreys, whom she had married in 1940. MMP, or $J \times J'$ as it is affectionately known, was first published in 1946, and after many editions and revisions, it was reprinted shortly before she died. It has educated generations of students and is still a recommended text for courses on mathematical methods in universities all over the world. Harold was knighted in 1953 for services to science, and Bertha became Lady Jeffreys (or Lady J as she was known to her students). Theirs was a long and happy marriage. Harold died in 1989 when he was almost 98.

Bertha always played a very active role at Girton, holding a large variety of offices at various times. She will probably be remembered most in her capacity of Director of Studies in Mathematics, a post she held from 1949 to 1969. The generations of women mathematicians she selected, advised and taught have gone on to propagate her influence to ever wider circles. She took a warm personal interest in all her students, and they were often invited to 'open house' on Sunday evenings at the Jeffreys residence.

It seemed inevitable that Bertha should be drawn into Harold's research interests. She was fluent in Russian and translated the book "Nutation and Forced Motion of the Earth" by E. P. Fedorov. In the 1970s she worked with Harold on editing his collected papers; these were very important pieces of work, for Harold was a pioneer not only of geophysics and seismology, but also of hydrodynamics, celestial mechanics and probability. Typical of Bertha's sense of humour, and of her desire to understand the origins of scientific knowledge, even down to the matter of notation, were two papers she wrote in her eighties, "A Q-rious tale ...", about the origins of the parameter Q used in electromagnetism.

Never a narrow specialist, Bertha was a skilful linguist, with a wide knowledge of literature, and music played a very important part in her life. She was an accomplished pianist and 'cellist. Her advice, on both professional and personal matters, was never stereotyped; she approached each problem with an open mind and an enormous amount of common sense. (One of her many gifts to College was the first washing machine for undergraduate use.)

In recognition of her leading role in education in the twentieth century, Bertha was awarded honorary doctorates by the University of Saskatchewan in 1995 and the Open University in 1996. During her retirement, she continued to be very involved in Girton, as a Life Fellow. After Harold's death, she travelled extensively, visiting friends as far away as Australia. She maintained a very active correspondence world-wide, often sorting out details of scientific history which she remembered better than most. Her ninetieth birthday lunch was attended by about 140 of her former pupils from several continents, a wonderful tribute to a very special teacher and friend. After her ninety-sixth birthday, her health, which had been amazingly good for a nonagenarian, gradually declined and she became less and less mobile. Right to the end, her mind was as sharp as ever; she died at home on 18 December 1999.

Olga Taussky-Todd

The daughter of an industrial chemist, Olga Taussky was born on August 30, 1906, in Olmutz in the Austro-Hungarian Empire (now Olomouc in the

Czech Republic). While she was still a child, the family moved to Vienna, and then, in the middle of the First World War, to Linz, where her father was director of a vinegar factory. Although her father would have preferred Olga and her sisters to go into the arts, he recognised Olga's mathematical ability and set her the task of working out how much water to add to various vinegars to attain the acidity level required by law. She rose to the challenge and her solution was put into daily use in the factory.

In 1925, Olga entered the University of Vienna to study mathematics. Gödel was a fellow student and friend there. She solved an important problem in number theory and obtained her doctorate in 1930. She then moved to Göttingen where she had a temporary position editing the volume of papers on number theory in Hilbert's Collected Works. Her job was very demanding and she regretted not having more time to interact with the large number of top mathematicians gathered there, who included Landau, Courant and, of course, Hilbert. By speaking out in her defence on one occasion, she became friends with Emmy Noether, one of the founders of modern algebra, who then ran a seminar in class field theory because of Olga's presence that year. Courant advised Olga not to return to Göttingen in the autumn of 1932, because of the growing political tension at the University, so she spent the next two academic years back at the University of Vienna, supplementing her small salary by tutoring.

While in Vienna, Olga applied for a research fellowship at Girton, which she had seen advertised in the newsletter of the International Federation of University Women. After sending in the application, she received an invitation to spend a year at Bryn Mawr, a particularly attractive offer because Emmy Noether was to be there. She accepted it, and then received the offer of a fellowship at Girton with "a very generous stipend of 330 pounds a year" and with "great freedom" to do what she liked. She was very pleased that Girton allowed her to spend that first year at Bryn Mawr. She chose to travel to the United States on a boat from Liverpool, thinking that the proportion of native English-speakers would be higher than on a boat from a port in southern England. She claimed that she acquired most of her English on that journey. The year at Bryn Mawr was a happy and stimulating one. She often accompanied Emmy Noether on her weekly trips to Princeton, making many friends and mathematical contacts there. Princeton was a "dream place" for her.

In June 1935, Olga arrived at Girton. She enjoyed the privileges of being a don, after student status at Bryn Mawr, but found it hard to fit into Cambridge mathematically; at the time, she was very interested in topological algebra and nobody in Cambridge was working in that area then. Nevertheless, she did some satisfying research in her first year, but then seemed to spend most of her second year applying for jobs, going to interviews and supervising students, partly to gain experience of teaching in English.

The next year saw Olga in London, with a junior position at one of the women's colleges in London University. This was not a particularly easy time: her teaching load was extremely heavy, her boss was unsympathetic and not all of her colleagues were very friendly. She wrote "They saw me as a foreigner. This had not been the case at Girton College where people had quite a liking for my foreign accent. Girton also had a scholarly attitude in everything...". However there was one overwhelmingly positive outcome of this time in London; at an inter-collegiate seminar, she met John (Jack) Todd, who had a similar position at another London college. They were soon working together on a mathematical problem and were married in 1938, the start of a very long and happy partnership.

The Todds moved eighteen times during the war. At first they lived with Jack's family in Belfast, while teaching at Queen's University. There Olga started working on matrix theory which subsequently formed a large part of her research activity. She then resumed teaching at her London college which had moved to Oxford to be safer from air raids. Then, on leave of absence from London University, she worked at the National Physical Laboratory at Teddington from 1943 to 1946.

In September 1947, Olga and Jack went to the United States as a result of an invitation to Jack to work on the exploitation of high speed computers at the National Bureau of Standards at its new field station in Los Angeles. While waiting for the facilities there to be finished, they spent time at the Institute for Advanced Study in Princeton in the group headed by von Neumann. Olga was very happy to be back in Princeton and the visit did much to restore her after the difficult war years. After a spell in Los Angeles and another year in London, the Todds moved to the National Bureau of Standards in Washington DC. Olga had the position of "consultant in mathematics", which involved a multitude of activities. These included refereeing papers, responding to cranks, inviting distinguished visitors and recruiting promising graduate students and postdocs as N.B.S. employees. She contributed number theory problems for the computers, and has been described as a "computer pioneer who provided significant contributions to solutions of problems associated with applications of computers".

Olga was ready to return to academic life when in 1957 she and Jack were both offered positions by California Institute of Technology, he a full professorship and she "a research position of equal academic rank ... with the permission but not the obligation to teach". The awkwardness of this position was probably a result of the fact that no woman had ever taught at Caltech before. Olga was pleased when she was given tenure in 1963, and even more pleased to be made a full professor in 1971, the first woman at Caltech with this academic rank. Her years at Caltech were extremely active and productive. She taught many graduate courses and established common research interests with many younger colleagues at Caltech and elsewhere, as well as with the more senior ones. She supervised more than a dozen Ph.D. students; one of them, Charles Johnson, wrote: "Olga Taussky-Todd often said that number theory was her first love, but in many ways she had the greater impact on her second love: matrix theory. She was involved with many of the major themes of twentieth century research in matrix theory, and the vast majority of her Ph.D. students were in matrix theory, several being major developers of the field in the latter half of the century. Perhaps most important she had an aesthetic sense and taste for topics that served to elevate the subject from a descriptive tool of applied mathematics or a by-product of other parts of mathematics to full status as a branch of mathematics laden with some of the deepest problems and emblematic of the interconnectedness of all of mathematics. Her influence on what people do and how well they do it will continue to be felt for some time" [1].

After retiring at the mandatory age of 70, Olga received the title of Professor Emeritus of Mathematics. She still supervised Ph.D. students for several years after this, continuing with her research and attending seminars until well into her eighties. Her death on October 7, 1995, at the age of 89, was a consequence of a broken hip from which she never fully recovered.

Olga wrote about 300 papers and her influence on many areas of mathematics is profound and long-lasting. She received many honours including the Gold Cross of Honour, First Class, the highest recognition given by the Austrian Government. She was a "Woman of the Year" for the Los Angeles Times in 1963; this gave her pleasure because it made Jack happy and could not make her (then all male) colleagues jealous.

Conclusion

Why were these women so effective and successful in their careers, with their varying backgrounds, two from the same English county, and one from far away in the Austro-Hungarian Empire? Their families had in common their understanding of the importance of education for women, and it seems to have been taken for granted that they would go to university and have careers. They were all very much in the minority as women mathematicians, but only Bertha Jeffreys found that she was excluded from some student societies because she was female. Two of them had supportive partners, while the third was free of domestic chores for most of her academic career because she lived in college. The only one to encounter the "two-body problem" was Olga Taussky-Todd, who for a while had to accept somewhat unsatisfactory positions in order to work in the same place as her husband. All three had a real love for mathematics and continued to take an active interest well into their retirements. They all maintained clear minds until the end, living until 89, 96 and 97; maybe mathematics really is a healthy occupation.

Acknowledgment

This article draws heavily on obituaries which I wrote for the Girton Review, and I acknowledge permission to include the material.

References

1. C. Johnson, *Not. Am. Mat. Soc.* **43**, 839 (1996).

WHAT ABOUT THE WOMEN NOW?

RUTH M. WILLIAMS

Department of Applied Mathematics and Theoretical Physics
Centre for Mathematical Sciences, Wilberforce Road
Cambridge CB3 0WA, England
and
Girton College, Cambridge CB3 0JG, England
rmw7@damtp.cam.ac.uk

Based on three studies, a survey describes some of the challenges faced by women mathematicians in Cambridge during the last decade of the twentieth century.

Keywords: Women mathematicians; University of Cambridge.

Introduction

The high-achieving women, described in the articles "And what became of the women?" [1] and "Three great Girton mathematicians" [2] seem far removed from the more typical Cambridge woman mathematics student, and a natural question to ask is how the milieu changed during the twentieth century, the lifetimes of the three Girtonians in focus. The percentage of women students did increase, but not by as much as one might have expected, given the fact that all the formerly all-male colleges started admitting women in the 1970's and 1980's. Girton, the oldest women's college in Cambridge, started admitting men in the late 1970's, and soon they were in the majority among the mathematics students. All the mixed colleges were eager to admit women mathematicians, but the numbers were usually rather small, and the new set-up sometimes led to problems of isolation and intimidation, which had not existed before in the generally supportive atmosphere of the women's colleges.

This article focuses particularly on the experience of Cambridge women mathematicians in the last decade of the twentieth century, as described in three reports. In the first case a university initiative gave impetus to a movement which, in some senses, was waiting to happen, in that a number of women in the Department of Applied Mathematics and Theoretical Physics (DAMTP) were feeling frustrated and undervalued, and they were encouraged to give voice to their concerns. In the second case, a mathematics teacher was given the opportunity to spend a term in Cambridge and see for herself how the mathematics course was functioning. The third case came again from a university initiative, this time bringing in investigators who worked in many faculties to try to understand the underperformance in women in examinations. The section of their report on the mathematics faculty is discussed in detail here.

As with all surveys which obtain material from a group of volunteers, there may be a bias in the comments made by a self-selected group. I quote a number of statements which give a rather extreme, and often negative, account, not because they are likely to be representative, but because they highlight the problems experienced by a small minority, and concern issues which still need to be addressed.

Women in DAMTP report

In 1991, heads of departments in Cambridge University received a circular from the Secretary General about equal opportunities for women, prompted by, amongst other things, the facts that 38.7 percent of undergraduates were women, whereas there were less than 10 percent of women in academic posts. With the encouragement of the Head of Department of DAMTP, the late David Crighton, a group of women in DAMTP, including research students, postdocs and staff, met a number of times and produced a report [3], which was also a reply to a University report of August 1991 on sexual harrassment. The report was not merely a response to the two university initiatives, but more an expression of the difficulties which the women felt in their professional lives, partly as a result of having to work in a predominately male environment.

The report first considered numbers. Cambridge was far out of line with other British universities over the number of women mathematicians admitted - only 16 percent in 1985-7, and increasing only very slowly since then, whereas most other universities admitted more than 25 percent women. Then once it had admitted them, Cambridge mathematics was losing women at every step of the academic ladder faster than it was losing men. This began with the disproportionately low number of firsts obtained by women which effectively disqualified most from going on to do research. The proportion of women research students was 12 percent, and women postdocs, college lecturers and temporary lecturers less than 10 percent. There was no woman with an established university post in DAMTP. The report pointed out that these dismal numbers were not an inevitable fact of life; for example, in 1991, over 20 percent of the two most senior grades of mathematician in the CNRS in France were women.

The next item of discussion was attitudes. Discounting the men who were always supportive and those who were so tiresome as to be lost causes, the women felt that amongst the majority of men, there were those who assumed wrongly that their customs and perceptions were held by women; in seminars questions tended to be more aggressive and confrontational, and men interrupted others much more than women did. Many women found this behaviour off-putting, and were intimidated by the pressure they felt to make any contribution intellectually fire-proof. The small amount of mainly mild sexual harrassment was felt by the women to be partly a consequence of their percived inferior status, and in some respects was seen to be less upsetting than being put down intellectually. Women who had been through the Cambridge system felt that, as undergraduates, they had sometimes been made to feel stupid by their male supervisors,[a] more than in the experience of their male peers.

A large number of suggestions of remedies formed the remaining, most substantial, part of the report. It was felt that a policy statement needed to be made, saying that the gender imbalance in the department was unsatisfactory and that action needed to be taken. The first arena for this was in undergraduate admissions; the reformed Tripos (the name of the Cambridge mathematics course) made it unnecessary for applicants to have Further Mathematics, a second mathematics A-level or school-leaving examination (its requirement previously had seemed to be a stumbling block for girls as only about 15 percent had been obtaining grade-A passes in it). It was also very important for women applicants to be interviewed appropriately. Secondly, in undergraduate teaching, some women should be giving lecture courses, partly as role models; there was a bold suggestion that there should be at least one woman lecturing a course in each year of the undergraduate Tripos. Women students should be encouraged to speak out if they were having problems with unsympathetic male supervisors. Thirdly, at the post-graduate level, able women should be strongly encouraged to do Part III Mathematics, an intensive year of preparation for research, which is essentially a prerequisite for being accepted as a research student in Cambridge. It was suggested that there should be at least two courses given by women

---

[a]In Cambridge, tutorials are called supervisions.

in Part III. Once they have been accepted to do research, women research students should be introduced into the department by someone supportive, and should be made to realise that they should seek help, rather than battling on in silence, if things were to go wrong in their research or in their relationship with their supervisor. Finally, at the postdoc level and beyond, the DAMTP women were not in favour of positive discrimination, but felt that women candidates should be actively encouraged to apply for posts. As a matter of good practice, there should be at least one woman on any appointments committtee, and consideration should be given to the effect that years of childbearing and raising have had on the careers of women applicants. Senior women in the department would feel less marginalised if they were more involved in both teaching and committee work, in spite of not being amongst the tenured staff. It was suggested that a mentor be appointed for each new postdoc (male or female) and subsequently a mentoring scheme for women was introduced, whereby a number of women staff, postdocs and research students were available to talk to any woman undergraduate or postgraduate about any matter of concern.

After the Women in DAMTP report was produced, the Head of Department sent it out to all members of DAMTP, with a letter emphasising his support and pointing out that problems were sometimes caused for women by careless speech or inappropriate jokes. An open meeting of the department to discuss the report took place in March 1993 and there was general support for the recommendations made. Subsequently, David Crighton facilitated the appointment of some women (and men) to posts of Assistant Director of Research.[b] Holders of such posts were eligible for promotion to established university positions, and such promotions did indeed happen in some cases.

The DAMTP women continued to meet until the middle of 1994 to discuss the implementation of the suggestions in their report. On a number of fronts, there was very little progress (for example, raising the number of women students), but there were some positive signs. For instance, a colloquium for school teachers had been geared more towards girls' schools; a booklet of welcome to the department had been prepared and was being used to help all newcomers to find their feet; several of the senior women had given lecture courses. The mentoring scheme had helped some research students, but had been used very little by undergraduates, perhaps because of lack of effective publicity.

---

[b]This was only possible because in most cases the colleges employing these people agreed to contribute substantially to the costs of their salaries.

Although it is not strictly relevant to the situation in Cambridge, it is perhaps appropriate to mention the founding of the Women in Mathematics Day, which is now held annually under the auspices of the London Mathematical Society. A speaker at one of the early meetings for the women in DAMTP was Dusa McDuff (mentioned in the article "And what became of the women?" [1] as one of the few women mathematician Fellows of the Royal Society). Dusa is now based in the USA, but was a research student in Cambridge, and had not found it at all easy to feel part of the Faculty. She not only encouraged the DAMTP women in their efforts but also suggested organising a meeting for women mathematicians from other universities, for mutual encouragement and also as an opportunity for women to give talks about their research in a supportive environment. It was originally called the "British Women in Mathematics Day" and took place in a number of different locations before being adopted by the LMS. It has been a source of inspiration and friendship for many women, particularly those who feel rather isolated in their departments.

Women and the Mathematical Tripos: Myth and Reality; the Salter Report

At around the time when DAMTP was considering the report by the women, another report was being written by Ruth Salter, who had spent the Michaelmas Term, 1993, as a School Teacher Fellow Commoner at Corpus Christi College. She had gone to Girton to read mathematics in 1950, and then taught mathematics in a variety of schools, including, at that time, an independent girls' school in London. During her term based in Cambridge, her aim was to investigate the transition from school to university mathematics, focusing particularly on the experience of women, and to see whether there was any truth in the myth that Cambridge was not a good place for young women to study mathematics. She proceeded in two ways, firstly by trying to put herself in the place of a first year student, attending lectures, doing problem sheets and sitting in on supervisions, and secondly by talking with mathematicians at all levels, undergraduates, postgraduates and teaching staff. Her report "Women and the Mathematical Tripos: Myth and Reality" [4] summarised what she had discovered and experienced.

Ruth attended the full set of first year lectures, attempting all the example sheets, and found the standard very high. She felt that great effort had been made to make the material accessible to students from weak school backgrounds, and that the view that the Cambridge course caters mainly for those who will get firsts is unfounded. She commented that some students from independent schools, who had been spoonfed, were finding the transition more difficult than some from state schools who had already had to learn how to organise their time. By the second year, there was a hint that for some people a concentration on learning how to do well in exams was overriding an earlier desire to understand and question.

The supervisions which Ruth attended seemed relaxed, with students willing to ask questions and responding well to stimulating teaching. However she met people whose experiences had been less positive, and, in particular, a number of women who had changed to another subject blamed unapproachable supervisors, who had made them feel inadequate and intimidated. Some suggested that women supervisors are sometimes more sensitive and confidence-inspiring to weaker students.

The encouragement, at that time, of applications to Cambridge from those who had studied only single subject mathematics A-level has already been mentioned in the context of the DAMTP report, and Ruth was interested in investigating how this was working. She found that the number of successful applicants was very small, and one woman student found that her supervisors and fellow students assumed that she could not possibly do well with such a limited background; she proved them wrong.

More generally, the attitudes met by women mathematics students were not always helpful, and were sometimes damaging. A male undergraduate told Ruth that the number of women mathematics students was quite good considering that "mathematics is not a girls' subject", and she found "more than a vague impression that some men assume that women are less able mathematically". She remarked that "women were more likely to be intimidated than stimulated by the male combative quality". Since the number of women overall was low, and Newnham, New Hall and Lucy Cavendish College (the only colleges taking only women students) had quite a number of them, the rest were spread quite thinly over the other colleges, and Ruth found sixteen cases of a sole women mathematician in her year in her college. Some thrived, but others felt uncomfortable and became more and more reticent. Not being confident enough to ask questions in supervisions made the situation even more difficult. If the number of women students were to increase, problems of isolation would not apply and gender issues would become less important. A number of women changed to other subjects because they felt undermined and discouraged in their first year. Ruth pointed out the need for more women supervisors and lecturers, both to avoid such problems and to provide positive role models.

Ruth left Cambridge convinced that, as a teacher, she should encourage able young women (and men) to apply to Cambridge, provided that they not only enjoyed mathematics, but also had real commitment and enthusiasm. The gender problems were not insurmountable, and she was reassured by the mentoring scheme which the women in DAMTP (as described earlier) were putting into place.

Indicators of Academic Performance

The continuing disquiet in the University of Cambridge about the discrepancy between the percentages of men and women obtaining first class results in the Tripos led to the establishment in 1996 of the Joint Committee on Academic Performance. The Committee recognised that although much research on this had already been done in faculties and departments, and by the Education Sub-Committee of the Senior Tutors' Committee, a more general university-wide investigation would be appropriate. Funding was obtained for a project entitled "Indicators of Academic Performance", to take place between April 1997 and October 2001. The qualitative research was done by Dr. Christine Mann and the quantitative work by Dr. Patrick Leman, together with a team of statisticians, and the final report [5], submitted in November 2001, was written by Chris Mann.

Although the project was initially set up to investigate the under-representation of women amongst those awarded firsts, its brief was then widened to a general study of the factors affecting examination performance. The method adopted was to look at the differences between groups categorized in different ways (gender, ethnicity etc) and then explore the reasons for the differences. The focus was on the cohort of students who started their undergraduate careers in 1997, and graduated in 2000 or 2001 (the so-called "millenium cohort"), and the factors considered were gender, social class, ethnicity, type of school attended and A-level results. (The A-level results were subsequently ignored because it turned out that over 90 per cent of the students had at least 3 A-grades at A-level.) In the cohort, the numbers were dividedly roughly evenly between men and women, and between independent school and state school background. The overall finding was that the class of degree awarded was correlated with subject studied, gender and ethnicity, but not with social class or type of school attended. Details of these findings are given in a report by Leman [6].

Even though the most significant factor affecting the percentage of firsts was the subject studied, I shall now focus on the conclusions about gender in the report; not unexpectedly, it was found that, proportionately, men get more firsts than women, and also more lower seconds and thirds. This was true even in subjects like English where women were not in the minority.

After collecting the initial data, the team of investigators interviewed students and staff in a number of subjects, to try to find out why there were such differences in performance. Chris Mann then continued in email contact with the 200 or so students from the cohort who were willing to be involved; this entailed questions asked by Chris and also spontaneous comments by the students.

Chris Mann's main conclusion from the general investigation concerned gender differences in the approach to learning, which seemed particularly strong in the physical sciences and mathematics. Women tended to see their course as an opportunity to increase their understanding and to focus on the subject area "in itself", so examination performance was a by-product of learning and personal development. On the other hand, men tended to be alert to "performance" aspects early on, with examination performance often the main target of learning. These differences affected preparation for examinations, with men concentrating more on examination technique, and women working hard to try to show the extent of their understanding. Women could suffer from trying too hard to do well, on the one hand, and from fear of failure on the other. Men tended to have more confidence in their innate ability and their skill in using examination techniques to good effect. These different attitudes could produce problems in supervisions when men wanted to move through all the problems, while women wanted to understand fundamental principles and could become inhibited if their questions were seen to hold up proceedings.

I now turn to the (rather substantial) part of the report addressing issues in the Mathematics Faculty. During the years 1990-2000, 35-40 percent of men had been awarded firsts, compared with 20-25 percent of women. These numbers already led to problems in fulfilment of the stated aim of the Cambridge Mathematical Tripos, to provide a challenging course suitable for students aiming to do research and those going into other careers. Possible reasons for the failure to do this included the fact that success in Tripos examinations was not necessarily good preparation for research, and, what is more relevant here, some potential researchers, women in particular, did not obtain high enough examination grades to be able to continue in mathematics.

At the undergraduate level, the intake in mathematics was roughly 25 percent women; just over 20 percent of the Part III students were women, and about 10 per cent of the research students. About 45 percent of the women were from single sex schools, and very few were from a lower middle class or working class background. The women who had taken STEP[c] as part of their entry requirement had, on average, significantly lower grades than the men. In fact STEP was seen as a deterrent for women, both because of a lower level of preparation and because of a reluctance to hold an offer for fear of failure.

In the millenium cohort of students, only a quarter of the mathematics students agreed to take part; of those, 40 percent were women, so women were actually over-represented. The students were asked to define excellence in mathematics, and the really able ones inderstood that this was not the same as excellence in Tripos examinations. For the first class students, the Tripos was seen as a sort of competition involving solving problems under time pressure, rather than showing other mathematical skills which they valued, in particular the creative reflective research mode. Those less successful at examinations included those who wanted to understand more deeply and not focus on problem solving techniques (with women often falling into this category). An "excellent" mathematics student was felt to be one with confidence, motivation, persistence and resilience. One male staff member viewed mathematics at research level as a very macho activity: "getting in there" and doing hard calculations, with success usually associated with men. One woman commented that "the expectation here is that you'll go at problems like a guided missile. I don't do that! I sit down and say 'oh, what a lovely problem!' And spread out feelers. That may be completely wrong here but doesn't diminish my mathematical ability. No - it does not." Some male staff members had no concept of what it would be like to be a good female mathematician; they had different expectations of what constituted good work by males and females.

The report high-lighted the effect of cultural context on achievement. The "maleness" of the Faculty, and the "unsettling thrill of being taught by the very best" sometimes proved overwhelming for the female minority. Some staff seemed overly concerned with their status outside the institute, and with indicators like college league tables, and so were distanced from students' real concerns. The gender imbalance meant that the Faculty was

---

[c]Most colleges in Cambridge make conditional offers in mathematics based not only on A-level grades but also on the Sixth Term Examination Papers (STEP), which are designed to differentiate between the large number of applicants who have A-grades in their A-levels. The STEP questions are intended to be more challenging than A-level ones, and to rely less on how well the student has been taught, but more on innate ability and potential.

dominated by a competitive ethos, in which some lecturers seemed to be trying to prove that they were cleverer than anyone else. The brusqueness of interactions with some male staff members and the lack of role models for women had negative effects. The attitude of male undergraduates to female lecturers was sometimes very critical; one commented that "there are two ways of thinking about women on the staff. Either they are there due to discrimination against men (and so may not be of the same standard) or they are there because they were better than men at the selection process. We have no way of telling which is which but when we are given step-by-step solutions to straightforward equations on a handout (so that she doesn't have to try to do it in front of us on the OHP) you do begin to wonder". It was very hard for women students to see a female role model treated in a demeaning way by male students. In the face of this type of behaviour, it was sometimes tempting for women to "go native" and try to function in a more male way, which could cause further problems.

To make a successful transition from school mathematics to the subject at university, students needed ability, love of the subject and, in most cases, hard work. Those not in the inner circle of participants in the Mathematical Olympiads (British and International) sometimes felt left out, and this could be particularly off-putting to women. Sometimes the very able were less mature, and established a working and social atmosphere which was inimical to women, who used words like arrogant, childish and competitive to describe it, and suffered from remarks that appeared to be made to put them down. Amongst the staff, there seemed to be an attitude that mathematical ability compensated for social ineptitute.

The gender balance meant that most supervisors were men and this sometimes contributed to the negative learning environment for those women who were less able, less well-prepared, lacked confidence or saw supervisions as a forum for asking questions. While there were many excellent male supervisors, there were still some who become aggressive or sarcastic when women got problems wrong. Women were sometimes humiliated when they make a mistake - one supervisor said "You may be doing mathematics but you are a girl - just a girl". One woman reported that she had asked a question in a supervision and was told that the answer was trivial; her male supervision partner asked an identical question some months later and was praised for his enquiring mind!

For students who lacked confidence, specific feedback could be very helpful. This could be from a supervisor or Director of Studies (the college official overseeing the academic aspects of a student's life) but was perhaps even more useful when it arose from a group of peers discussing their work. In this context, it was interesting that although women in their first year were sometimes anxious not to be singled out by being supervised with other women, by their third year women often admitted to being more comfortable having a female supervision partner. The withholding of feedback, sometimes as a protective measure, could make women feel that the college thought they did not really fit into the system. Support mechanisms for students who were really struggling were not well-developed in some colleges.

So who was thriving in the Mathematics Tripos? The course was very effective at stretching the brightest to their full potential, but maybe not so good at meeting the needs of a wider range of students, some of whom eventually decided to change subjects. Of course some who changed had come to Cambridge with the intention of going on to study computer science, physics, or economics, say, but others changed because they realised that Cambridge mathematics was not for them, and a larger proportion of women than men were in this category. The reasons for this included the expectation in some quarters that women were no good at mathematics, bad preparation for the course and the pace of teaching in the first year, which could make it seem too hard straightaway and give rise to a fear of mathematics. Struggling women sometimes tried to work even harder and this soon reached a point of being counter-productive. Even able women sometimes left before Part III, often because they did not enjoy the highly competitive atmosphere, an "alien environment" according to one staff member. Women often wanted to feel that the socially able and those with wider interests were as welcome in the Faculty as the more typical mathematics student. However it was pointed out that it was a self-perpetuating system, with no incentive to change.

The Report concluded with a number of specific recommendations. Firstly, there should be more senior women employed in the Faculty, to provide positive role models for women students, to give women colleagues the opportunity to experience a culture where women's achievements are encouraged and celebrated, and to give access to an alternative learning environment with which women students might feel more comfortable. Secondly, the school-university transition should be made easier by a system of mentoring starting even before students' arrival in Cambridge, regular peer group meetings within colleges to discuss work, 'drop-in' academic help centres run by established students and finally the availability of sessions with lecturers for students finding the material difficult. On the teaching front, there should be training in basic lecturing skills and compulsory training for supervisors which should include material on gender dynamics. Finally, the Faculty should make it clear in all advertising material whether it intends to continue with its current gender balance, competitive ethos and its focus on the top 30 percent of students admitted.

The Report was discussed in detail by the Mathematics Faculty Board, and comments were sent to the relevant university committees. It was felt that the final recommendation in the report was rather unfair, as the Faculty was constantly reviewing and sometimes reforming the Tripos in an attempt to cater for the wide range of ability of the students. There were also continuing efforts to try to improve the gender balance, and steps were being taken to try to make women students feel comfortable in the Faculty. This revolved mainly round social events where students could interact with senior women and with their female peers. The Faculty Board questioned the implication of the report that the underperformance of women was a result of the teaching environment in Cambridge, when in fact the Tripos results were consistent with the STEP results, indicating that the problems existed before the women students even reached Cambridge. A great deal of self-study material for STEP preparation was provided by the Faculty and there were plans to start a short residential course in the Easter vacation for students with no access to help with STEP preparation at their schools. It was felt that the different styles of learning discussed in the report were not so clearly divided along gender lines, but were more a function of confidence levels in both men and women, and that the examinations were appropriate for both styles. The negative comments on supervisions were obviously a cause for concern, but it was felt strongly that they were not typical, because of the self-selecting nature of the group of students monitored. Supervision training sessions were now run jointly with the University Staff Development team, and attendance was compulsory for new research students.

Conclusions

In these three reports, the same points have been made repeatedly: the problems for women mathematicians of being a minority in a predominantly male culture, the lack of role models, and the difficulties caused by insensitive attitudes and remarks, often leading to the undermining of confidence. It is useful, and perhaps alarming, to ask how much has changed in Cambridge mathematics as a result, directly or indirectly, of these reports.

The numbers have improved at the senior level in DAMTP, where there are now one woman Professor, two Readers, one Lecturer and an Assistant Director of Research. In the Department of Pure Mathematics and Mathematical Statistics (DPMMS), there is currently no woman with an established post in Pure Mathematics but there is one Senior Lecturer in the Statistical Laboratory. There are at least eight women College Teaching Officers in mathematics. In DAMTP, the proportion of women postdocs and research associates is about 18 percent, of research students 20 percent and of Part III students 17 percent. These represent improvements over the figures in the early nineties, apart from in the case of Part III students (where the percentage in DPMMS is even lower). The proportion of women undergraduates in the Faculty has not changed significantly in the last ten years.

The activities arranged specifically for women students have fallen off since the days when the DAMTP women were meeting regularly; the attendance at parties for women undergraduates became rather low, and Part III women did not seem very interested in organised lunches after the early days of their course, but the senior women still meet occasionally. The mentoring scheme has lapsed. On the positive side, partly as a result of all the work of Marj Batchelor to create a more supportive environment for Part III students, the general atmosphere among the graduates is much more friendly and the women seem to feel more confident and much less isolated. There is now a Graduate Mathematics Society, which organises very successful events, academic and social, and the women are mainly very well-integrated into this.

Even though there are more women giving lectures, there are not enough for all students to be lectured to by at least one women each year, when sabbaticals and other commitments are taken into account. The compulsory supervision training for research students is very worth-while, but the problems of insensitivity and inappropriate teaching style sometimes arise with supervisors who are well-established and very unlikely to be willing to go to a training session. Many students still find the transition to university mathematics vey challenging, and although there are certainly male students who lack confidence, this does seem to be a more common problem with women.

Overall, there has been a lot of progress, but while there are still women (and men) students whose experience of Cambridge mathematics is difficult and unnecessarily uncomfortable, there need to be support systems in place, even if they are used infrequently. In particular, a mentoring system could provide a lifeline in some circumstances. Gatherings arranged specifically for women mathematicians may be felt by many to be unnecessary or even offensive to men, but can be an invaluable opportunity to share experiences and a source of encouragement, particularly for those feeling vulnerable in the predominantly male culture of mathematics.

Acknowledgment

I thank Caroline Series for encouraging me to write this article, Ruth Salter for providing a copy of her report, and Anne Davis and Ann Mobbs for providing material relevant to the other two reports.

References

1. Caroline Series, And what became of the women?, this volume (2009).
2. Ruth M. Williams, Three great Girton mathematicians, this volume (2009).
3. Vivien Chamberlain et al., Women in DAMTP, private circulation (1993).
4. Ruth Salter, Women and the Mathematical Tripos: Myth and Reality, private circulation (1993).
5. Christine Mann, Indicators of Academic Performance, Cambridge University (2001).
6. Patrick Leman, The role of subject area, gender, ethnicity and school background in the degree results of Cambridge University undergraduates, *The Curriculum Journal* **10**, 231 (1999).

MATHEMATICS IN SOCIETY (TAKING INTO ACCOUNT GENDER-ASPECTS) — A ONE-SEMESTER COURSE (BSc)

C. SCHARLACH

Institut für Mathematik, MA 8-3, Technische Universität Berlin
10623 Berlin, Germany
schar@math.tu-berlin.de
www.eecs.tu-berlin.de/eecs/prof_christine_scharlach/

We report on a one-semester course with the same title taught at the Humboldt University to Berlin in the winter semester 2006/2007. We believe that the mixture of topics of this course as well as the teaching methods are a good and very efficient way to fill some gaps in many universities mathematics (and mathematics education) curricula. Since many of the inner capacities (psycho-social skills) cannot be taught as subjects, they must rather be modelled and promoted as part of learning.

Keywords: Mathematics; society; gender; curricula; inner capacities; beliefs.

1. Introduction

What is mathematics? A student of mathematics should deliberate about mathematics, e.g. about questions like:

- What are my beliefs about mathematics?
- What is the role of mathematics in society?
- Why should people learn mathematics?
- Where is mathematics used and for what?

We report on a one-semester course taught at the Humboldt University to Berlin in the winter semester 2006/2007. We believe that the mixture of topics of this course as well as the teaching methods are a good and very efficient way to fill some gaps in many universities mathematics (and mathematics education) curricula. Since many of the inner capacities (psychosocial skills) cannot be taught as subjects, they must rather be modelled and promoted as part of learning (UNESCO, Education for All) [1]. The main goal of the course is to start and support a process of reflection. The students were supposed to get a look on mathematics 'from above' and to reflect upon their beliefs [2] about mathematics. Topics were for example: maths as a profession, maths between the humanities and sciences, maths and gender or maths in history, philosophy and politics. Only a small part consisted of lectures, a bigger part was organized as a seminar, also we had professional mathematicians as guests for interviews or as co-lecturers, and we went on excursions. The seminar was led together with a graduate student in gender studies. Accompanying to the course were exercise groups to impart the necessary working skills (in part from the humanities). We had 14 participating students, 9 female and 5 male. Eight of them were diploma students in the 2nd year, three were (advanced) teacher students and three had a background in gender studies. Six seminar talks were presented.

2. Studying math at German universities - the current situation

We mainly describe the bachelor (and master) program [3] at the TU Berlin, which is best known to us. Since we taught at the HU Berlin, we know the mathematics diploma program (and the program for teacher students) there. Furthermore we compared with bachelor and master programs at TU München and U Hamburg. Because of the Bologna process the situation at other German universities should be comparable.

Since the start of the Bologna process in 1999 the German Higher Education system changed to match the performance of the best performing systems in the world, notably the United States and Asia. One of the priorities of the Bologna process is the introduction of the three cycle system (bachelor/master/doctorate) [4]. In comparison with the former diploma system the curriculum is tightly organised and students (and teachers) have less freedom of choice. A student who studies mathematics at a German mathematics department takes classes in mathematics (about 75-80 % ot the total credit points), some classes in a second subject (about 10-15 %), usually some field of application of mathematics like computer science, physics or economics. Due to the way universities are organized there is often little exchange between the different subjects, and even between the different special fields of mathematics. Thus a student gets highly specialized knowledge, but it is difficult to get a broader picture, a look on mathematics "from above". We cite from the introduction of the script "Geschichte der Mathematik I" (History of mathematics I) of W. Hein [5]:

"An der Universität lernt man Mathematik, indem man verschiedene logisch strukturierte und sorgfältig organisierte einzelne mathematische Theorien kennenlernt, die zudem meistens beziehungslos nebeneinander stehen."

(At the university one learns mathematics by getting introduced to different logically structured and carefully organized separate mathematical theories, which are moreover mainly not connected to each other.)

To prepare the students for a professional life students have to take some additional classes. The focus hereby is different in the mathematics departments, but commonly there are hardly any offers made directly by the mathematics departments [6–8]. At the TU Berlin, students can choose from all classes offered at the university (about 5 % of the total amount of credit points in bachelor and specialised master programs, 23 % in the general mathematics master program). This seems to be similar at the TU München, here is a choice provided which is called "überfachliche Grundlagen" (interdisciplinary basics). Bachelor students at the U Hamburg can choose classes from a selection "general professional qualifying skills" (15 %), so far there is no master program.

3. The one-semester course at HU Berlin

3.1. *Mathematics as a profession*

First we studied mathematics as a profession by reading and presenting parts of the first chapter of "Viewegs Berufs- und Karriereplaner" [9] (profession and career planner). The first chapter "Warum Mathematik studieren?" (Why study mathematics?) gives a broad overview over different areas of mathematics which are relevant for professional careers as a mathematician. The other chapters of the book are informative, too. We left it to the students to continue after having them introduced to the book. To study the subject in a broader perspective, including historical developments in the last century and the gender aspect, we continued with the book "Traumjob Mathematik! Berufswege von Frauen und Männern in der Mathematik" [10] (Dream job Mathematics! Career paths of women and men in mathematics) by Abele, Neunzert and Tobies. This book is interesting in several aspects, we will mention only some of them. In the introduction the authors state eight opinions of preconceptions about male and female mathematicians, which they investigate in the following chapters.

Examples are:

Vorurteil 1: Mathematiker sind weltfremd und wenig sozial, sie beziehen ihre Zufriedenheit aus der Arbeit und schöpfen nicht aus sozialen Beziehungen. Mathematik ist "unweiblich" und "wider die Natur der Frau".

(**Preconception 1**: Mathematicians are ivory-tower and unsocial, they draw their satisfaction out of work and not from social relations. Mathematics is "unwomanly" and "against female nature".)

This preconception takes up some stereotypes about mathematicians. To reflect upon these stereotypes and more general the beliefs about mathematics is a main goal of the course and was treated in several ways later on.

Vorurteil 6: Wenn Frauen sich für Mathematik interessieren, dann wählen sie in erster Linie einen Lehramtsstudiengang.

(**Preconception 6**: If women are interested in mathematics, then they choose to study the teaching profession.)

It is easier to test this preconception for truth since it can be analysed by quantitative methods (which is done in the book, cp. p. 162f.). Anyhow we did not resolve the question about the content of truth in these preconceptions directly. At this point of the class it just was a way to start the reflection process.

To gain more knowledge about the career paths of women and men in mathematics we continued with the historical part: For the first half of 20th century there are both quantitative and qualitative (biographical) studies about alumni of teacher training (Chap. 3.1), of graduate studies (Chap. 6.1) and of diploma programs (only qualitative, Chap. 4.1). Again, we did not study the whole book, which treats in detail the current situation. Further studies were left to the interested students. Only Chapter 8, in which the preconceptions stated at the beginning are analysed, was content of the class, as part of the evaluation process at the end of the semester. (In a further course we gave a report on the data in Chap. 2 which gives an overview on the quantitative development of the studies of mathematics in Germany from 1925 resp. 1970 to 2000. This seemed to be overwhelming for the students and we recommend to present just some current data from the own university or country. Often students are not aware of the fact that still there are remarkable differences in the participation of women and men in mathematics.)

It turned out that the diploma students had only vague ideas about their professional goals after leaving university. This is rather typical for German students in mathematics. Their decision to study mathematics is led by their interest (and talent) in mathematics and the many different professional opportunities. Often this is different for teacher students, their professional goal is set. Anyhow, mathematics as a profession should be part of their professional knowledge to counsel their students. Another way to reflect about mathematics as a profession is to meet mathematicians.

3.2. *Interviews with mathematicians*

The course was accompanied by a series of interviews with mathematicians. The main purpose is the presentation of role models. But obviously there is a lot more about getting to know some mathematicians, their work and their professional life, and to talk with them about their beliefs about mathematics. We either visited them at their working place or they visited us. Beforehand the students got some information, usually the vita of the interviewed persons and some material about their work (their homepage or the homepage of their company, articles presenting them and their work, articles written by themselves about their work or mathematics in general; cp. the web presentation of the course [11]). The interview partners were informed beforehand about the special interests of the students (the subjects they presented in the seminar part) and were asked to prepare for questions about their personal experiences concerning:

- role models (family, teacher, others),
- career path (school, university, others),
- beliefs about mathematics,
- importance of mathematics and their math education in their current life,
- mathematics in the society,
- who does how mathematics,
- mathematics and gender.

We were lucky that a wide variety of mathematicians (people with a university degree in mathematics) accepted our invitation. Four of our interview partners are female and the other four are male, all of them established in their professional fields. We had guests from all three Berlin universities, active in mathematics and mathematics education. Three of them are professors, one is a scientific assistant on a permanent position and one a women's representative. Also from the educational professions, a high school teacher (Studiendirektor) visited us. Furthermore we visited a project manager of Siemens and an officer of the "Gesamtverband der Deutschen Versicherungswirtschaft (GDV)" (German Insurance Association). Most of the interviews were led by the teacher of the course, but also the students asked many questions. One of the interviews started with a presentation given by the interviewed person. The small group made a very open and personal atmosphere possible. Another interview was special, consisting mainly of a lecture about the analysis of a mathematical text [12] (by Emmy Noether), presented in two parts by the interview partners. The idea and the main contents of this lecture came from the interviewed person. That we had a balanced number of female and male guests for the interviews partially was luck, but the gender issue was present in most parts of the course and also a topic by itself.

3.3. *History and philosophy of mathematics*

Questions about the foundations of mathematics, about the truth/proof-problem (are proof and truth the same in mathematics?) or the existence of mathematical objects are not treated in a standard mathematics curriculum. Most people believe that mathematical knowledge is independent of experience (a priori by Kant) and therefore absolutly reliable. It is based on proofs and thus one can not argue about mathematical results. These beliefs are rarely challenged at the university. Thus we had planned in advance that the fundamental controversies over the epistemology of mathematics (Grundlagenkrise) at the beginning of the 20th century should be treated in the course. An easy introduction can be given by the SWR2-radio play "Eine Menge stelle ich mir vor wie einen Abgrund" [13] (I imagine a set as an abyss) by Kai Petersen. But it is difficult to give a short overview. It turned out that two students had studied these problems before and were willing to present an introductional overview into the philosophy of mathematics as a workshop in the seminar part of the course. Another student already had some knowledge about Gödel and his incompleteness theorems and gave a presentation in which he included the radio play mentioned before. To round this up we organized the last seminar session as a "role-playing game", a fictive conference meeting about "Perspektiven der Grundlagenforschung der Mathematik" (Perspectives of the research on the foundations of mathematics). The students assumed the historical roles of Gödel and Quine (platonism), Hilbert (formalism), Brouwer (intuitionism),

Russell (logicism) and due to the lack of a historical role model Pieper-Seier (mathematics and gender). The students prepared for the conference with parts of Chap. 2 of Bettina Heintz, "Die Innenwelt der Mathematik" [14], in particular pp. 38–41 (Platonismus, Chap. 2.1.1), pp. 47–52 (Formalismus), pp. 52–55 (Chap. 2.2), pp. 55–60 (Platonismus, Chap. 2.2.1), pp. 60–69 (Grundlagenkrise); and an interview with Irene Pieper-Seier, "So ist es, so macht man das und das ist objektiv und ganz genau?" [15], by Klupsch and Günzel. Furthermore we suggested to them an overview article by Gerald Walti, "Eine kurze Einführung in die Philosophie der Mathematik" [16].

3.4. *Gender meets Mathematics*

Part of the course (the "seminar", two hours per week) was developed and taught in team work with Daniela Döring[a] (MSc in cultural sciences), a PhD-student in Gender studies. Since only three students had a background in gender studies (no student was enrolled in gender studies) we introduced the subject with two articles:

(1) "Warum Gender-Studies?" [17] (Why gender studies?) by Christina von Braun,
(2) "Feministische Naturwissenschaftskritik. Eine Einführung" [18] (Feminist natural science criticism. An introduction) by Dorit Heinsohn.

The first article gives an introduction to gender studies (Geschlechter-forschung, in Germany) in general. Our focus for the course was

- on the definition of "Geschlecht" (in the German language this is the only word for gender and sex) and its relation to "Natur" (nature),
- on the change of gender roles and models at the beginning of the 20th century,
- reasons for this change,
- the definition of "Gemeinschaftskörper".

The second article describes the development of the research area "Feministische Naturwissenschaftskritik (FNWK)", its main questions and problems and how they can be sorted as three dimensions of FNWK (following E. Fox Keller [19]: Women in Science, Science of Gender and Gender in Science).

---

[a]http://www2.hu-berlin.de/gkgeschlecht/kolleg/ddoering.php

Here the focus was

- the goals and questions of FNWK,
- the meaning of "Gender",
- the three dimensions of FNWK,
- the meaning of "soziale Unternehmen",
- how can natural science and science criticism meet and what might be the problems,
- are long lasting intensive controversies in natural sciences possible.

Before we discussed the articles in class, the students had to read (part of) them, led by the questions (see above) we gave them. It turned out that most of the students found the articles interesting but hard to read. Part of the reason might be that as students of mathematics they are not used to this type of reading (which is from our point of view an argument against early specialisation). What we had not expected was that the article of Heinsohn let to an intensive argument about how to acchieve gender equity in the use of German language. Even though there are official regulations and recommendations since at least 1990 (cp. the collection of links on a website [20] of the GenderKompetenzZentrum), they still have not reached academic life (and daily life). Daniela Döring, the co-lecturer, took the opportunity to spontaneously include a newspaper article ("Das Eva-Braun-Prinzip" von Thea Dorn, TAZ, 29.11.2006) to further stimulate the discussion. This part of the course ended with excerpts of the book "Die Innenwelt der Mathematik" [14] (The inner world of mathematics) by Bettina Heintz (Introduction, Chap. 1 and Chap. 7.4). Based on this text we mainly discussed the question if a mathematician must anchor her or his work in society. Before we continue by describing some contents of the course curricula concerning the scientific community in mathematics, we would like to state a view point on mathematics which the students newly formulated at the end of the "Gender meets mathematic"– sessions. The students said that one could see mathematicians as members of a faith community i.e. mathematics is a form of religion.

3.5. *The scientific community in mathematics*

We had two main strategies to introduce the students to the scientific community in mathematics (apart from the interviews already mentioned). The practical one consisted of visits of scientific talks/events. Having made several suggestions we agreed with the students on the following ones:

- "Was zählt. Präsenz und Ordnungsangebote von Zahlen im Mittelalter"[b] (What counts. The Presence and Medial Functions of Numbers in the Middle Ages), Internationale Arbeitstagung, Berlin, Helmholtz-Zentrum fr Kulturtechnik, November 2006.
- Berlin Mathematical School (BMS), Kovalevskaya Lecture, "Congruences for the number of rational points of varieties defined over finite fields"[c] by Hélène Esnault,
- (voluntarily) BMS Friday colloquium, "Gödel's Vienna"[d] by Karl Sigmund,
- Ringvorlesung "Geschlecht in Wissenskulturen", HU Berlin, "Das Geschlecht natürlicher Zahlen. Zum Zusammenhang von Zählen und Zeugen"[e] (The gender of natural numbers) by Ellen Harlizius-Klück,
- (voluntarily) Ringvorlesung "Literarische Inszenierungen naturwissenschaftlichen Wissens", TU Berlin, "Ein Jahrhundert der Mathematik. Dilettantische Betrachtungen über Harsdörffer/Schwenters 'Deliciae Physico-Mathematicae' "[f] by Martin Disselkamp.

On a more theoretical level we studied the question "Was ist ein mathematischer Beweis? (Konkurrenz unter Mathematikern)" (What is a mathematical proof? (Competition between mathematicians)) by reading and discussing the article "Manifold Destiny. A legendary problem and the battle over who solved it" [21] by Sylvia Nasar and David Gruber. In addition two students presented the biografies of Grigori Perelman and Shing-Tung Yau. (As an exercise one student looked for the stereotypes about mathematicians presented in the article, another one summed up what is said about the proof-validation-process. The lecturer sketched the relations between the mathematicians mentioned in the article.) More information about the mathematical background can be found in an article of the sciences magazine "Breakthrough ot the year: The Poincaré Conjecture–Proved" [22] by Dana Mackenzie.

3.6. *What is mathematics?*

We used more techniques to deliberate about mathematics and our questions asked in the beginning. Creative techniques like mind mapping,

---

[b]http://www2.hu-berlin.de/kulturtechnik/bsz.php?show=veranstaltungen&which=waszaehlt&page=expose&lang=en
[c]http://www.math-berlin.de/BMS-Friday3.pdf
[d]http://www.math-berlin.de/BMS-Friday5.pdf
[e]http://www2.hu-berlin.de/gkgeschlecht/downloads/rv_wise0607.pdf
[f]http://www.literaturbaum.de/index10.html

vernissage, visualisation or autobiographies were used. We watched the movie π, USA 1998, Darren Aronofsky (author and director). And scientific methods like seminar talks, presentations and work shops given by students were used. To complete this report we will state the subjects the students chose:

- Introduction to Philosophy of Mathematics,
- Two Cultures, Math between sciences and the humanities,
- Foundational crisis of mathematics and Gödel,
- The roman and the sumerian numbers,
- Mathematics and politics: "German mathematics",
- Mathematics as a profession.

References

1. UNESCO, The six goals of education for all, goal 3 – promote learning and skills for young people and adults (02.10.2007), `http://portal.unesco.org/education/en/ev.php-URL_ID=41579&URL_DO=DO_TOPIC&URL_SECTION=201.html`.
2. Katja Maaß, Veränderungen der schüler(innen)vorstellungen über mathematik durch modellierungsprobleme im unterricht – erste ergebnisse einer empirischen studie, in *Beiträge zum Mathematikunterricht* (Franzbecker, Hildesheim, 2003), pp. 429–432.
3. TUB, Studienordnung für den bachelorstudiengang mathematik an der technischen universitaet berlin (Version 24.01.2006), `http://www.math.tu-berlin.de/~studber/download/StuPO-Bachelor/Bach_M_StO.pdf`.
4. European Commission, The Bologna process. towards the European higher education area (13.08.2007), `http://ec.europa.eu/education/policies/educ/bologna/bologna_en.html`.
5. W. Hein, Warum mathematikgeschichte studieren? Aus der einleitung zum vorlesungsskript geschichte der mathematik i (Version 25.9.2008), `http://www.uni-siegen.de/fb6/geschmath/vorwort.html?lang=de`.
6. TUB, Alle module der bachelor- und masterstudiengänge mathematik, technomathematik und wirtschaftsmathematik (Version 12.02.2009), `http://www.math.tu-berlin.de/~studber/ModuleMath.shtml`.
7. TUM, Modulhandbuch des studiengangs bachelor mathematik (Version 01.10.2008), `http://www.ma.tum.de/Studium/Modul`.
8. UHH, Lehrveranstaltungen (Version 10.02.2009), `http://www.math.uni-hamburg.de/teaching/lectures/index.html`.
9. *Vieweg Berufs- und Karriereplaner* (Vieweg, Braunschweig, Wiesbaden, 2001).
10. A. Abele *et al.*, *Traumjob Mathematik!* (Birkhaeuser, Basel, 2004).
11. C. Scharlach, Mathematik in der gesellschaft (unter berücksichtigung von genderaspekten) (Web presentation of the course, Version Summer 2007), `http://www.mathematik.hu-berlin.de/~schar`.

12. E. Noether, Idealtheorie in ringbereichen, *Math. Ann.* **83**, 124 (1921).
13. K. Petersen, Eine menge stelle ich mir vor wie einen abgrund, die grundlagenkrise des mathematischen denkens (19.02.2000), `http://db.swr.de/upload/manuskriptdienst/wissen/wi000219.txt`.
14. B. Heintz, *Die Innenwelt der Mathematik: Zur Kultur und Praxis einer beweisenden Disziplin* (Springer, Wien, 2000).
15. R. Klupsch and S. Günzel, So ist es, so macht man das und das ist objektiv und ganz genau - interview mit irene pieper-seier, in *KORYPHÄE*, Vol. 12, pp. 52–55, 1992.
16. G. Walti, Eine kurze einführung in die philosophie der mathematik (Nov. 2001), `http://www.ifi.uzh.ch/groups/ailab/teaching/NAISemi01/Presentations/Philosophie_Mathematik.htm`.
17. C. von Braun, Warum gender-studies? vortrag anlässlich der feierlichen eröffnung des studiengangs gender-studies (21. Oktober 1997), `http://edoc.hu-berlin.de/docviews/abstract.php?id=6008` [Version 12.02.2009].
18. D. Heinsohn, *Feministische Naturwissenschaftsforschung. Eine Einführung*, in *Feministische Naturwissenschaftsforschung: science & fiction*, eds. B. Petersen and B. Mauss, Schriftenreihe NUT - Frauen in Naturwissenschaft und Technik, Vol. 5 (Talheimer Verl., Mössingen-Talheim, 1998), Mössingen-Talheim, pp. 14–32..
19. E. F. Keller, *Origin, History, and Politics of the Subject Called 'Gender and Science' - A first Person Account*, in *Handbook of Sciences and Technology Studies*, eds. S. J. *et al.* (SAGE Publications, Thousand Oaks, London, New Delhi, 1995), Thousand Oaks, London, New Delhi, pp. 80–94.
20. HUBerlin, Genderkompetenzzentrum: Aspekte sprache (Version 04.09.2008), `http://www.genderkompetenz.info/genderkompetenz/handlungsfelder/sprache/aspekte/`.
21. S. Nasar and D. Gruber, *Manifold Destiny. A legendary problem and the battle over who solved it.* (The New Yorker, 28.08.2006), `http://www.newyorker.com/archive/2006/08/28/060828fa_fact2`.
22. D. Mackenzie, *Breakthrough of the year: The Poincaré Conjecture–Proved* (Science magazine, 22.12.2006), `http://www.sciencemag.org/cgi/content/full/314/5807/1848`.